THE

OXFORD PORTLAND CEMENT COMPANY LIMITED

SITUATED AT

KIRTLINGTON & SHIPTON-ON-CHERWELL CEMENT WORKS

OXFORDSHIRE

COMPANY HISTORY, RAILWAY & LOCOMOTIVES

Frontispiece: A great view of the impressive Shipton-on-Cherwell, Oxfordshire, cement works towards the end of its working life and showing the three exchange sidings next to the Oxford to Banbury former GWR main line, that can be seen just peeping in on the extreme right.
Courtesy Ian Cuthbertson.

THE

OXFORD PORTLAND CEMENT COMPANY LIMITED

SITUATED AT

KIRTLINGTON & SHIPTON-ON-CHERWELL CEMENT WORKS

OXFORDSHIRE

COMPANY HISTORY, RAILWAY & LOCOMOTIVES

by

COLIN JUDGE AND CHRIS DOWN

INDUSTRIAL RAILWAY SOCIETY

ISBN 95781912995 14 1
www.irsociety.co.uk

© Published in 2024 and distributed by the
Industrial Railway Society under the imprint
IRS Publications, 24 Dulverton Road, Melton Mowbray, Leicestershire, LE13 0SF.

Book prepared for printing by Karen Robertson for her father, Colin Judge.
Printed by Artisan Litho Ltd., Abingdon, Oxfordshire.

Dedicated to Geoff Cryer, Andrew Smith, and Huw Williams of the Industrial Railway Society for all their help and guidance – thank you, much appreciated.

Author's note: Some of the images are captioned as 'Author's collection' and have been obtained from international auction sites and they have had no information with them. Therefore, no copyright infringement within this book or use of photographs or images is intended.

Front cover photographs:
Top: Locomotives AB 2041/1937 and P 1378/1914 in the quarry area with P 1378 pushing the five wagon load of stone towards the crusher, in March 1969. *Courtesy Great Western Trust collection, Didcot.*

Bottom: Locomotives No. 3 and No. 6 alongside the conveyor tunnel in June 1967.
Courtesy Geoff Broadhurst.

Rear cover photographs:
Top: Locomotives P 1378/1914 – WESTMINSTER No. 5 and RSH 7742/1952 – C.F.S. No. 6 alongside the coaling and watering site in the quarry.
Courtesy Industrial Railway Society, photographer Gordon Green.

Bottom: A publicity photograph of the AtW No. 110 locomotive GOLIATH – airbrushed and inscribed for its potential customer – Oxford & Shipton Cement Ltd. *Courtesy Industrial Railway Society.*

Title page: Locomotive ALBION II built by William Bagnall (works number WB 2178 /1921) outside the Shipton-on-Cherwell locomotive shed on the 19th of March 1967.
Courtesy Industrial Railway Society; photographer Malcolm Briam.

Abbreviations used in this book:

O/S	Ordnance Survey.
OPCC	The Oxford Portland Cement Co. Ltd.
OCC	Oxfordshire County Council.
GWR	Great Western Railway.
w/n	Works number.
BCI	Blue Circle Industries PLC.
APCM	Associated Portland Cement Manufacturers Limited.
IRS	Industrial Railway Society.
ILS	Industrial Locomotive Society.

Contents

Acknowledgments

Prime Source – Courtesy National Archives; ref – BT 31/17576/86120 and MT 29/85.

***To my co-author* – Chris Down** for use of the huge amount of material from his personal files on the Alpha Cement Company, Blue Circle Industries PLC., plus his drawings, photographs, and numerous articles from The Times newspaper – and above all his personal support in the numerous writings he compiled for inclusion in the book.

Oxford History Centre of the Oxfordshire County Council Record Office for extracts from the Oxford Times and Oxford Mail newspapers (all from Microfilm).

Helen Drury, photographic manager of the Oxfordshire County Council History Record Office.

Geoff Cryer, for his computer work on the many tricky photographs and graphics.

Huw Williams for editing and preparing the text for publishing.

Chris Turner, Chris Potts, Tony Cooke for all the GWR and signalling information.

Professor Thomas Russell for the superb cement flow diagram.

Danielle Czerkaszyn of the Oxford University Natural History Museum for images and information on the Kirtlington quarry.

Dylan Moore, David Baird, Tom Burnham, Richard Sams, Raymond Jupp, Keith Maltas, Philip Cartwright, Jonathan Toyn and John Oxford for all the additional information on the history of the two cement works.

Simone Neat – Cherwell District Council.

Kevin Boys – former Blue Circle surveyor for Shipton information.

Industrial Locomotive Society – Stewart Liles – photographic officer, searching the photographs.

Industrial Railway Society – Andrew Smith for supplying the immense photographic selection, the Hunslet Engineering GA drawing and his continued support for this book from the Industrial Railway Society.

Industrial Locomotive Society – Chairman, Russell Wear for the information and help in the compiling of the Chapter on the Locomotives.

The Earthmovers magazine – Graham Black editor.

Keith Haddock (Edmonton, Canada) for his immense knowledge on the Excavators plus all the photographs, brochures and plans he supplied from his vast collection.

Martin Loader, Steve Banks, Robin Waywell, Kevin Lane, John Hutchings, Ian Cuthbertson, John Wolley, Murray Liston, Adrian Nicholls, Gordon Edgar and Geoff Broadhurst – all for supplying many of the photographic images.

Justin Edwards for all the information and photographs of the Ruston Excavators.

Laurence Waters for photographs and timetables from the Great Western Trust collection, Didcot.

Dr Michael Stansfield, Archivist for New College Oxford library for maps and documents.

English Heritage, Swindon for the aerial photograph of Shipton-on-Cherwell plant.

Robert Cowling, editor of Waterways World Magazine for the OPCC canal barge image.

Staffordshire Record Office – Liz Street for the drawings of the W. Bagnall locomotive.

National Library of Scotland for supplying some of the Ordnance Survey maps.

Kilmartin Group for their precis of the site planning statement, after closure of the Shipton Works.

Some of the reference books referred to, etc:

Kirtlington – An Oxfordshire Village by Vanadia S. Humphries.

The Early Years of the Oxford Cement Industry by C. Dosworth.

Victoria County History of Oxfordshire, VI (1959).

The Oxford Canal by H. J. Compton.

History of Kirtlington, Oxfordshire by J. McMaster.

Cement Manufacture by UK Companies, 1914-1994 by Peter Jackson.

Oxford Portland Cement Co. Ltd. by H. J. Compton.

Blue Circle – 50th Anniversary Booklet: OXFORD 1929-79.

Introduction

The thought of writing a book on a cement company and its works had never crossed my mind until, whilst searching for something else at the Oxfordshire County Council, Oxford History Centre Record Office, St Lukes Church, Cowley, Oxford, revealed an entry in their photographic records of twenty-five images, on Plate glass negatives, dated 1930 of the Shipton-on-Cherwell cement works. To make it more intriguing, there were no 'thumbnail prints' shown against the accession numbers, possibly meaning that these negatives had never been printed or ever used.

Looking at the relevant OS maps of the area, it showed that there was quite an extensive internal industrial railway system on the final site at Shipton-on-Cherwell and so this prompted me to ask questions to the Industrial Railway Society's photographic officer, as to how many industrial locomotives did this company own? The answer surprisingly came back as eleven, which is rather a large amount for just one middle-size company.

I was intrigued to find out more and so then enquired – if so, how many images does the Industrial Railway Society photographic collection hold of this company's locomotives and yet again the answer was even more astonishing – sixty-five – not to be out done, the Industrial Locomotive Society then offered me thirty-eight more. I must at this point also mention and thank the good size colour photographic images from the Great Western Railway Trust at Didcot, Geoff Broadhurst's private visit colour slides and so many more photographers that offered their photographs to be included in the book. Once all these were received and subsequently seen, it was time to get started to research the history in our local and national public records offices and then to choose which photographs to include.

There was an additional positive advantage to writing this book, as the IRS had within its membership, Chris Down, who is recognised as a specialist in the history of cement works. He has written articles and books on the subject and agreed to help me with my research, finally to become my co-author! Lastly, this book would not have been published without the enormous help of Geoff Cryer, whose computer wizardry has, after many hours of work, made the photographs, figures, drawings and plans ready to be reproduced to a very high standard – thank you Geoff!

My first task was to source the actual name of the company, which appeared to be The Oxford Portland Cement Co. Ltd. (OPCC) and again, where in the surrounding part of Oxfordshire were their works situated?

However, during my initial research of the OPCC, it appeared that previously there had been another earlier small cement works near to the village of Kirtlington, entitled the Kirtlington Cement Co (the site was often locally called Washford Pits). Following on from these small beginnings in cement production, a new, larger, and more modern company (the OPCC) was formed and situated roughly in the same area, again near to the village of Kirtlington. The original plant was constructed in 1906/07 and built within an existing stone quarry, alongside the all-important life-blood of the Oxford Canal.

Some twenty-two or so years later, the company's raw material resources (i.e. its stone quarry) ceased to be viable and so the company was moved to a new, larger works and huge quarry site at Shipton-on-Cherwell that the OPCC had had the foresight to have previously purchased some years earlier. This quarry and site were situated alongside the GWR Oxford to Birmingham main line.

Now this new Shipton-on-Cherwell Cement works, especially having later in its life, a large white chimney, had in the past been quite important to my own experiences, as I knew it as 'Smokey Joes'.

It was the last aerial turning point on the landing circuit to runway 19L at Kidlington airfield (now London Kidlington airport). During my training to be a private pilot and attending the local airfield at Kidlington for my course, you would hear every time you were committed to land on runway 19L, the words, something like "*turn over the white chimney at 1000ft – call finals – and descend to runway 19L*". Even sometimes on take-off the words would ring out "*climb to 1000 ft and turn either left or right over the large white chimney – set course*" or even given a compass reading to set you on your way once over this chimney.

So you can see that Shipton-on-Cherwell cement works was extremely valuable, not only to the cement industry, but also to pilots, especially when the weather had quickly deteriorated. That tall (250 feet) white chimney that you could see from miles away was a god-send to inform you where you were and certainly saved my life a couple of times and probably many other lives as well.

The story and history of this local company, OPCC (whose ownership and name has later changed several times) is fascinating from the beginning to the main cement works closure in 1986/87 and then to its complete demise in 1993. Its hidden importance has been taken for granted but is, and has been, important component of the building industry, not only locally but at later dates to the whole of the UK, especially during the two World Wars.

And now to the last and most important point.

I must record my many heartfelt thanks and acknowledgment to the enormous help from my co- author Chris Down who joined me towards the later part of the detailed research and preparation of this book. It became abundantly clear that the breadth and depth of this complex cement subject, could be immensely increased and improved historically by taking Chris on board; as I have stated earlier, he is recognised from the many articles and books he has written as a specialist on cement works and, particularly, their internal railway systems. Without his knowledge, assistance and the many reams of paper I read through in his collection, mostly from the records of the Blue Circle Industries PLC, I (or should I say 'we') would not have been able to complete the work – so thank you Chris!

Colin Judge 2024.

Plate 1: This fine view is taken as a pilot would be approaching runway 01 at Kidlington airfield, but you can see 'Smokey Joes', the 250 foot high white chimney of Shipton cement works prominently in the middle of the photograph, which, if landing from the other end on runway 19L, you can see that the white chimney would be the turning point from the aircrafts downward leg to line up for the cross wind leg before turning 'finals' to land on runway19L. A great visual help to all!

Author's collection obtained from an international auction site and reproduced under the Creative Common licence agreement.

Chapter One

What is PORTLAND CEMENT and its HISTORY?

Before the reader is introduced to the main topic of this book, being the history of the Oxford Portland Cement Co. Ltd. and its future site and owners, it is briefly worth considering – what is portland cement and where did the name come from?

Portland cement is relatively new in terms of the word cement, which has been around for thousands of years as an agent for sticking together stones and bricks to form a rigid structure. It is probably the most common type of cement in existence today and used almost everywhere in the world as one of the basic ingredients of concrete etc. It was developed in Britain from basic limestone in the early part of the 1800s by Joseph Aspdin, a Leeds bricklayer. The fine powder of portland cement is made from heating limestone and clay in an object called a kiln, to form a substance called 'clinker'*. This is then milled and a small amount of gypsum and other materials are added. There are various types available today, which is basically grey, but white portland cement is often used on buildings with that colour and type of stone. The whiteness is achieved by using purer materials, which gives the cement its white colour.

Plate 2: William Aspdin.

The name 'portland cement' was coined by Joseph Aspdin, (a born and bred Yorkshire man), from his patent in Leeds in 1824 and is derived from its appearance like the Portland Stone that was and is still available in Dorset, at the large quarries on the Isle of Portland. The substance that Joseph had made is now thought not to be true portland cement and, in fact, true portland cement was first made by his son William.

William Aspdin (the son of Joseph) 1815 – 1864 is credited as the real pioneer and inventor of today's portland cement, after his successful experiments in the 1840s.

William was born in Leeds and known locally as an 'incorrigible liar and swindler.' The second son of Joseph, he had joined his father's cement company in 1829 – leaving the company in 1841 to set up his own cement plant at Rotherhithe in 1842. He found that improving his father's cement formula by increasing the limestone content and burning it at a higher heat he could obtain a slower setting rate and a higher strength cement, suitable for use in concrete.

He did not take out a patent on this discovery but disguised the result to his employees by scattering coloured crystals into the kilns when they were ready for firing, saying that this mixture was the secret ingredient to his new discovery!

*Clinker makes up more than 90% of the portland cement, with a small amount of calcium sulphate (gypsum), which controls the setting time and up to 5% of other materials, which including clay, shale, sand, iron ore, bauxite, fly ash and slag. When the kiln is coal fired, the ash from the coal becomes a secondary raw material.
Clinkers are nodules with diameters of 1/5th inch, produced when the raw materials are heated to a very high temperature, normally in a kiln. This distinguishes portland cement from other forms of cement.

One of his rivals, Isaac Johnson (1811 to 1911) in 1845, at J.B. White & Co's plant at Swanscombe Works (opened in 1825/26 by James Frost, making a cement called 'Roman cement'), also made a similar discovery. William went into partnership with his eldest son James and bought a cement plant at Northfleet Creek, Kent (which came to be known as Robins Works) and operated from there in 1846.

He then moved to County Durham as Aspdin, Ord & Co but soon, in 1857, sold again and moved to Germany. There he built Altona and Lagerdorf cement works, which were the first commercial plants to make portland cement outside of Great Britain.

The Northfleet works was taken over by APCM (Blue Circle) in 1900 and they shut it down soon afterwards. APCM also bought the Gateshead plant, from the then owners Robins & Co. Ltd., shutting it down in 1911.

William was made bankrupt twice, fuelling the assertion that he was an inept businessman, a liar and dishonest; even a con-man and forger. He had had no formal training in chemistry and his 'inventions' appear to have been the result of fortunate accidents. One of his false claims was that his father Joseph's cement was used by Isambard Brunel on repairs to the Thames Tunnel in London. William died near Hamburg, Germany in 1864.

The so called 'modern' portland cement really came into being around 1756, when civil engineer, John Smeaton experimented with different limestones and other material substances, for use in the construction of his Smeaton's Tower lighthouse.

The reason for this was that he was commissioned to build the third Eddystone lighthouse. The first (1698-1703) and the second (1709-1755) had both been destroyed by the harsh elements. This new lighthouse was to be situated on a tiny rock, some 14 km from shore, which was submerged at high tide. He designed an ingenious structure of huge interlocking granite blocks with dovetailed joints which needed a water-proof mortar to seal the joints. After serious experimentation, he opted for Blue Lias limestone from Watchet in Somerset and pozzolan from Naples to make the mortar. This lighthouse remained in use until 1876, when it was replaced, since it rocked in the high winds because of the erosion of the rock it was standing on. The upper part of the lighthouse was re-erected on Plymouth Hoe, for all to see today!

Further portland cement development had been achieved in 1796 by James Parker on Roman cement, which was the popular cement at the time. However, in 1811 an Englishman Edgar Dobbs, patented a cement that was improved in 1817/18 by a Frenchman, Louis Vicat, who had chemical training and the resultant product was possibly the forerunner of today's portland cement.

In the 1850s to 1860s, the French imported portland cement mainly from England, to make concrete to build harbours at Cherbourg, Dieppe, and Brest. Later in 1853, at Pas de Calais they began their own limited portland cement production.

Other private developments continued until 1859, when John Grant, employed by the Metropolitan Board of Works, London drew up specification for portland cement to be used in the construction of large sewers in London. This became the 'standard specification' for the future portland cement manufacture.

The next step was the introduction in 1885 of the rotary kiln, which had been patented by Frederick Ransome and allowed better mixing of all the ingredients. The rotary kiln was further developed by making the kiln 'endless' which controlled the combustion rate perfectly and produced a better grade of clinker.

The setting and then hardening of portland cement was like other types of cement in use today. When the cement powder is mixed with water, it causes a complex series of chemical reactions, still only partly understood. The ingredients slowly crystallise, then they interlock, giving it the strength. After the initial setting, immersion in warm water speeds up the rate of setting, it was found that adding small amounts of gypsum prevented rapid 'flash' setting. Controlling the rate at which the cement sets means that portland cement can be used for many of the mixes in use today. Thus, setting times can be controlled depending on the use to which the cement is to be put; for mortar, plaster, grout, screed etc.

Today the cement industry is heavily scrutinised due to its environmental impact and various methods to reduce this impact are being investigated. The main environmental concerns that are associated with cement production include: airborne dust emissions, gases released during the combustion processes, noise and vibration from heavy machinery and the volumes of fuel consumed. It is estimated that portland cement contributes to about 10% of the world's carbon dioxide emissions and research is ongoing to replace portland cement production by what are termed 'supplementary cementitious' materials.

The main concerns are that portland cement is caustic, which will cause chemical burns when handled, severe irritation and even lung cancer from chromium ions found in the cement. But, on the plus side, the cement kilns of today are burning high levels of domestic waste products to help heat the kilns to the high temperatures needed. Many of the old worked-out quarries are being returned to nature or regenerating the countryside, so protecting our ever-decreasing natural world.

Note: The name PORTLAND used in this book's text refers to a material or process and is not a proper noun, so therefore is not capitalised.

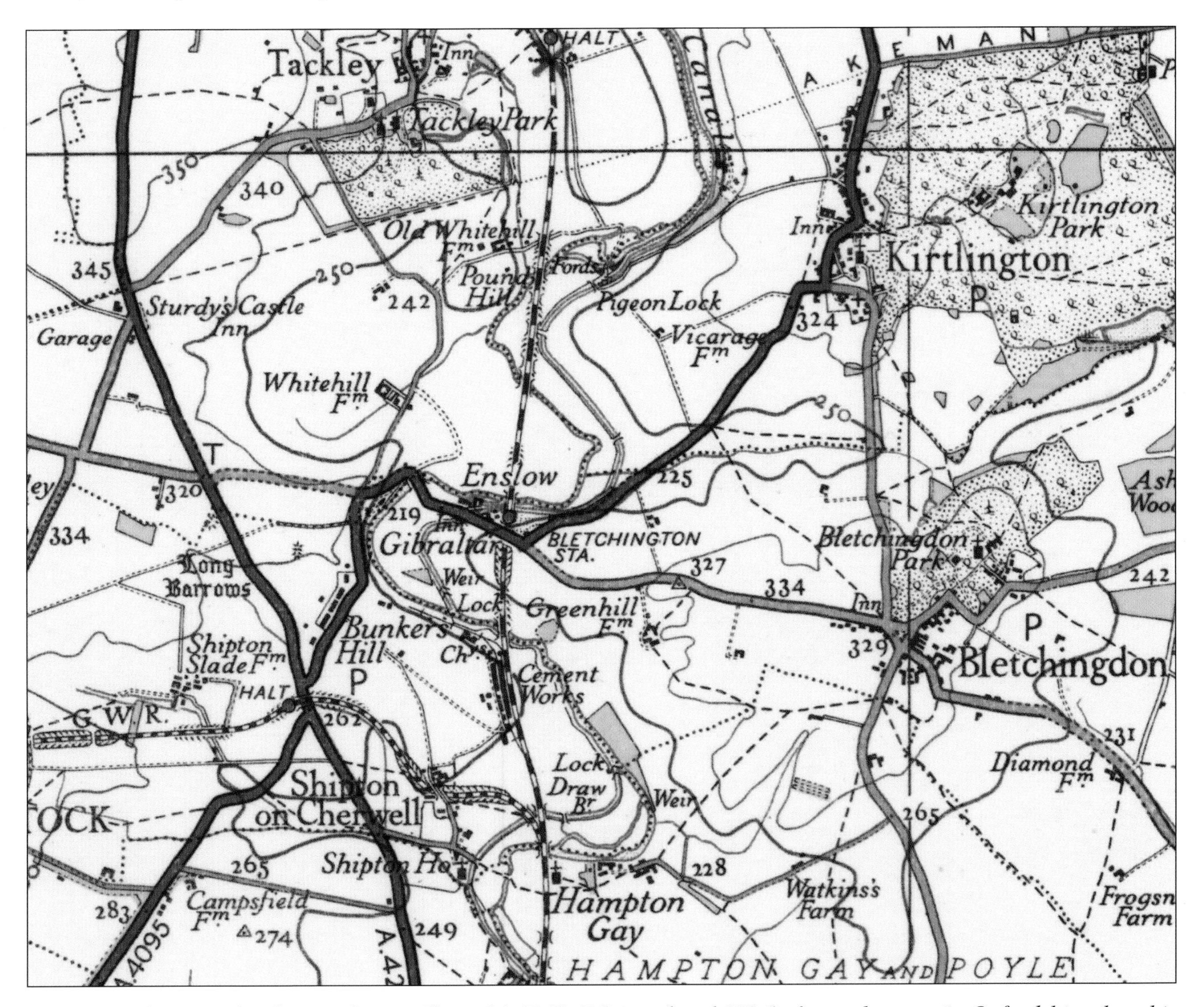

Figure 1: This map (Ordnance Survey Sheet 94, Fifth Edition dated 1936) shows the area in Oxfordshire that this book covers.

Reproduced courtesy National Library of Scotland under the Creative Commons Attribution Licence.

Chapter Two

Early Beginning at KIRTLINGTON of the OXFORD PORTLAND CEMENT COMPANY LIMITED.

Before delving into the history of the **Oxford Portland Cement Co. Ltd.**, whose works were situated at Kirtlington and Shipton on Cherwell in the Cherwell Valley, Oxfordshire, a short history of the Kirtlington quarry (grid reference SP494 199) is appropriate, as it is of national interest.

Small scale quarrying had taken place at Kirtlington since at least 1425, and continued through the Middle Ages, before the first Oxford Portland Cement Co. Ltd. works was constructed.

The quarry is listed as one of the most important geological Middle Jurassic vertebrate localities in the world and dates from around 166 million years ago. This area is part of the Forest Marble Formation sediments, where numerous important species of extinct mammals, including fossilised bones of the 15-metre long Cetiosaurus, a tooth of a Megalosaurus plus teeth of flying reptiles called pterosaurs, remains of sharks, Long Snouted Crocodiles and parts of a large marine reptile called a Plesiosaur have been discovered. In 1910, the first geologist was allowed into the quarry by the cement company to study in detail and record the various layers of materials in the limestone of the quarry face. The area also provides evidence of much more recent geological events, when the Ice Age was active and powerful rivers flowed through the area. The rocks that outcrop here are very good cement production and there are many disused quarries in the area to remind us of its important industrial past.

Plate 3: A superb photograph taken in 1923 of the long east face of Kirtlington Quarry and showing the narrow gauge tramway that used hand pushed trucks from the quarry face to the cement works crushers and mixers. The poles and wires are a mystery but their use could be for carrying the electric cables and lights for night working in the quarry? The various levels of stone and clay etc. on the quarry face can clearly be seen.

Courtesy Oxford University Museum of Natural History.

Only in 1974 was the Kirtlington Mammal bed thoroughly investigated revealing the many fossils as listed above, plus those of fishes, frogs, lizards, salamanders, turtles, making the 3.1 hectares (7.7 acre quarry) a geological Site of Special Scientific Interest, which is now part of the 7.4 hectares (18 acres) Kirtlington quarry Local Nature Reserve. All these finds were deposited millions of years ago when Oxfordshire was submerged under a warm tropical sea.

The value of the Oxfordshire minerals had long been recognised especially those of the Jurassic strata. There was a large deposit about half a mile from Kirtlington Village and small limestone quarries were founded in the area called Washford Pits, which had been worked on and off since medieval times. However, in 1875, near to the transport link of the important Oxford canal, a small cement company – The Kirtlington Cement Co – was established by Messrs. Lamprey of Banbury.

Later, in 1905, a local dignitary, Arthur Henry Dillon from Kirtlington Manor House, later Lord Dillon of Ditchley Park, Oxfordshire (a stockbroker in London), with Arthur John McMillan joined together to form a company. They managed to obtain a lease on the land they needed from Sir George John Egerton Dashwood, a local dignitary.

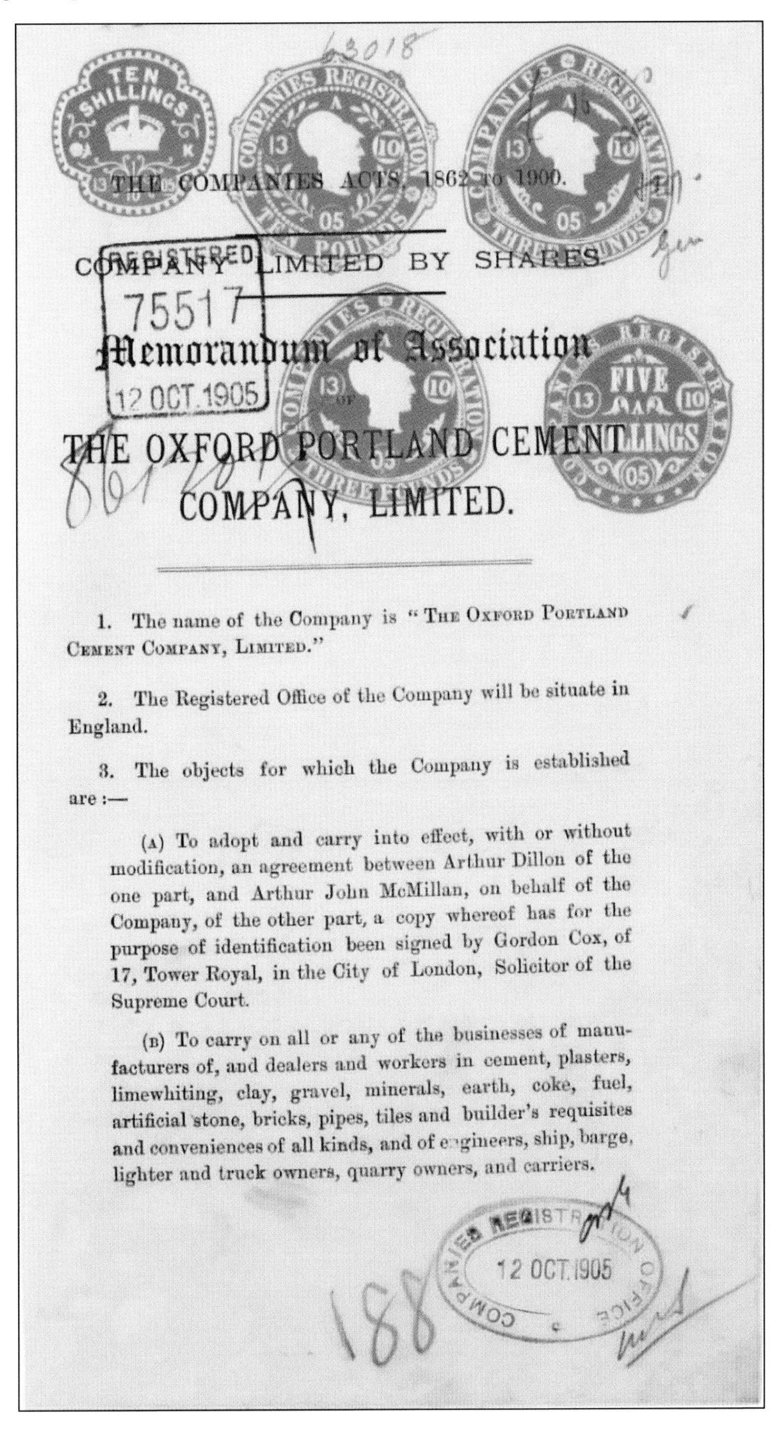

THE COMPANIES ACTS, 1862 TO 1900.

COMPANY LIMITED BY SHARES.

REGISTERED 75517 12 OCT. 1905

Memorandum of Association

OF

THE OXFORD PORTLAND CEMENT COMPANY, LIMITED.

1. The name of the Company is "THE OXFORD PORTLAND CEMENT COMPANY, LIMITED."

2. The Registered Office of the Company will be situate in England.

3. The objects for which the Company is established are:—

(A) To adopt and carry into effect, with or without modification, an agreement between Arthur Dillon of the one part, and Arthur John McMillan, on behalf of the Company, of the other part, a copy whereof has for the purpose of identification been signed by Gordon Cox, of 17, Tower Royal, in the City of London, Solicitor of the Supreme Court.

(B) To carry on all or any of the businesses of manufacturers of, and dealers and workers in cement, plasters, limewhiting, clay, gravel, minerals, earth, coke, fuel, artificial stone, bricks, pipes, tiles and builder's requisites and conveniences of all kinds, and of engineers, ship, barge, lighter and truck owners, quarry owners, and carriers.

COMPANIES REGISTRATION OFFICE 12 OCT. 1905

Plate 4: The original agreement in 1905 between A. Dillon and A.J. McMillan to form the Oxford Portland Cement Company Ltd. under the Companies Act of 1863. Courtesy National Archives; ref – BT 31/17576/86120.

Later three other interested parties also joined the two company founders; William Long Franklin, a builder from Deddington, Mark Morrell, a mining engineer from Clifton Hampden and John Lamb Spoor, a cement manufacturer from Rochester in Kent. Together these five promoted the Oxford Portland Cement Co. Ltd. The five had very different interests and it is not clear how this partnership ever really got together.

The initial agreement between A.H. Dillon* and Arthur John McMillan and with Sir George John Egerton Dashwood for the leasing of land from the Kirtlington Park Estate is shown in *Plate 6*. This document also includes the initial building agreement between Arthur Dillon and John Lamb Spoor and is dated 31st August 1905 (registered on 31 July 1906).

A.H. Dillon's mother, Ellen was Sir H.W. Dashwood's daughter and that family connection doubtlessly helped A.H. Dillon, and with McMillan obtaining from Sir George John Egerton Dashwood (H.W. Dashwood's son) a lease over approximately 193 acres of his Kirtlington Park Estate. The land was required for the construction of the new Oxford Portland Cement Company Ltd.'s works;' the lease being dated on 25th December 1905, valid for sixty years. The same agreement, dated in pencil for the 31st August 1905, states that John Lamb Spoor was to oversee the design, erection and equipping the works, when complete, with the most modern equipment. As appointed contractor, Spoor was to complete the works for the sum of £20,000, paid in instalments and to work two months at the plant and produce cement before the works were accepted and the final payments made – a novel way to ensure due diligence in his work! He was also to give advice and share his experience of cement manufacture and the organisation of the business. He also agreed to be a Director, but he disappeared from all documentation by 30th August 1907, including the directors registered list, being replaced by George Todd Brown, 5 South Parade, York – listed as a 'Gentleman' and as an 'additional director'.

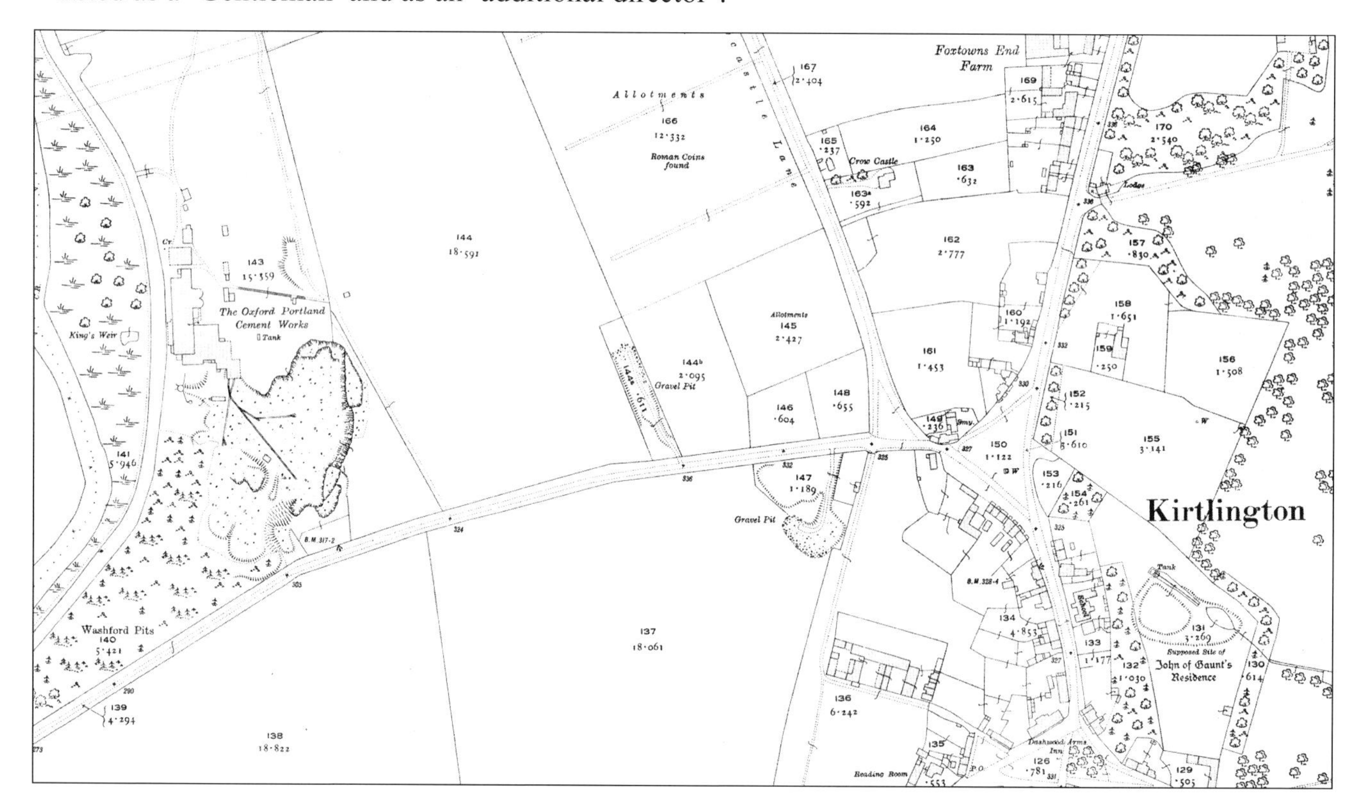

Figure 2: The 1922 OS map showing the area where the Kirtlington/Washford Pits cement works was built, to the west of the village of Kirtlington. Note the internal tramway tracks in the quarry.
Courtesy National Library of Scotland; Oxfordshire XX11.14 Kirtlington, Tackley etc. 1922.

* Arthur Henry Dillon, 18th Viscount of Costello-Galin (1875-1934) was the son of Hon. Conrad Dillon and Ellen Louisa Dashwood. He was a JP and High Sherriff of Oxfordshire 1931-1932. His wife was Hilda Dillon, nee Brunner and in the early days they resided at Steeple Aston, later Kirtlington Manor and finally at Ditchley Park when he was Lord Dillon.

Plate 5: The following three legal documents show the agreement for leasing the land for the cement works at Kirtlington and the building contractor agreement dated 31st of August 1905.

Courtesy National Archives; ref – BT 31/17576/86120.

86120 68426 This may be registered 30.7.06

FIVE SHILLINGS

SIX PENCE

REGISTERED 64471 31 JUL 1906

An Agreement made the Thirty-first day of August one thousand nine hundred and five **Between** Arthur Dillon of 47 Oakley Street Chelsea in the County of London Stockbroker of the one part and Arthur John McMillan of 17 Tower Royal Cannon Street in the City of London Esquire for and on behalf of the Oxford Portland Cement Company Limited (hereinafter called "the Company") of the other part **Whereas** these presents are supplemental to an Agreement dated the Thirty-first day of August one thousand nine hundred and five and made between Sir George John Egerton Dashwood of the one part and the said Arthur Dillon of the other part whereby the said Sir George John Egerton Dashwood agreed to grant a lease for a term of Sixty years from the twenty fifth December one thousand nine hundred and five of certain land and liberties for the getting of limestone and clay and for the manufacture of Portland Cement and otherwise in respect of certain land at Kirtlington in the County of Oxford known as the Washford Pits Limestone Quarries at the rents and royalties and upon the terms mentioned in the said Agreement for a Lease **And whereas** these presents are also supplemental to an Agreement (hereinafter called "the Building Agreement") dated the second day of August one thousand nine hundred and five and made between the said Arthur Dillon of the one part and John Lamb Spoor of the other part whereby it was agreed that subject to the conditions of capital of the Company being subscribed in cash to an amount of not less than Twenty thousand pounds and of the Company proceeding to allotment before the first September one thousand nine hundred and six the said John Lamb Spoor should erect and supply works for the manufacture of Portland Cement at Washford Pits aforesaid with all proper and necessary buildings machinery plant and apparatus and should carry on the business of the manufacture of Portland Cement at the said premises for a period of two calendar months and on the termination of that period hand over the said business to the said Arthur Dillon as a going concern with a staff of skilled workmen and foremen and that the said John Lamb Spoor should subscribe and pay in cash for Two thousand pounds of the ordinary share capital of the Company

REGISTRATION 31 JUL 1906

and would after the handing over of the works as a going concern to the said Arthur Dillon assist him in the organisation of the business as Adviser or as a Director of the Company for a period of three years and subject as aforesaid it was further agreed that as consideration the said Arthur Dillon should pay to the said John Lamb Spoor the sum of Twenty thousand pounds by the instalments therein mentioned. And that the said Arthur Dillon should be entitled to assign the benefit of the Building Agreement to the Company and that on execution by the Company of a deed to which the said John Lamb Spoor should be a party and by which the Company should adopt the Building Agreement and agree to perform and observe its covenants and obligations in the place of the said Arthur Dillon the said Arthur Dillon should be relieved from all further liability in respect thereof **And whereas** the Company has been formed principally with the object of adopting and carrying into effect with or without modification this Agreement **And whereas** the nominal capital of the Company is Fifty thousand pounds divided into Fifty thousand Shares of One pound each **Now it is hereby agreed** as follows:—

1 **The** said Arthur Dillon will assign to the Company the benefit of the recited agreement for a Lease and the benefit of the Building Agreement

2 **The** consideration for the said assignment shall be the sum of Two thousand pounds to be paid and satisfied by the allotment to the said Arthur Dillon of Two thousand Shares in the Company of One pound each credited as fully paid

3 **The** purchase shall be completed within one month of the first general allotment of the Companys Shares at the Office No. 17 Tower Royal Cannon Street E.C. of Messrs Cosc & Lafone when the consideration aforesaid shall be paid and allotted and thereupon the said Arthur Dillon and all other necessary parties if any shall at the expense of the Company execute and do all assignments assurances and things for vesting the benefit of the recited agreements in the Company and giving to it the full benefit of this agreement as shall be reasonably required.

4 **The** Company shall on completion execute a deed to which the said Arthur Dillon and the said John Lamb Spoor shall be parties by which the Company shall adopt the Building Agreement and agree to perform and observe its covenants and obligations in the place of the said Arthur Dillon.

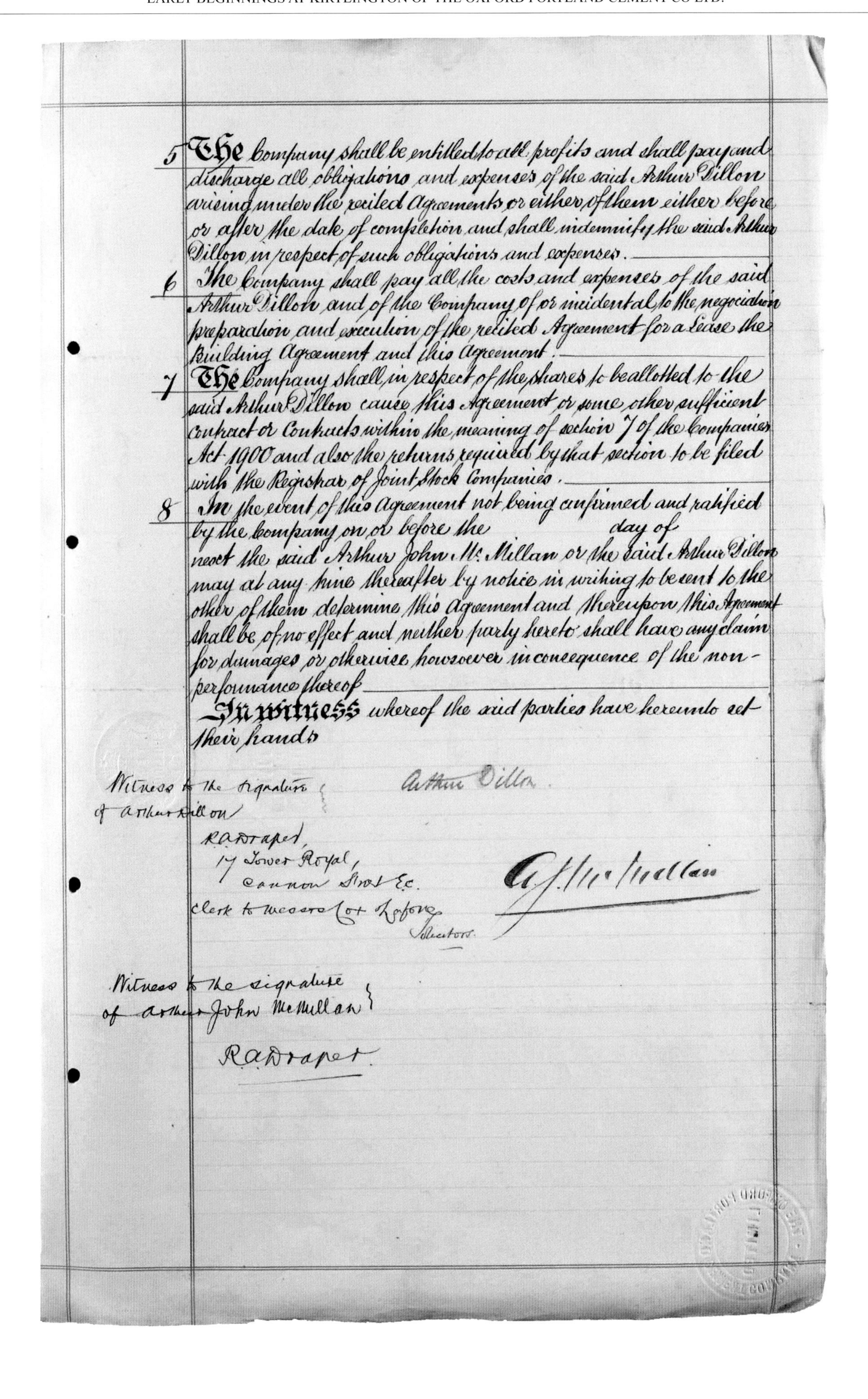
5 The Company shall be entitled to all profits and shall pay and discharge all obligations and expenses of the said Arthur Dillon arising under the recited Agreements or either of them either before or after the date of completion and shall indemnify the said Arthur Dillon in respect of such obligations and expenses.

6 The Company shall pay all the costs and expenses of the said Arthur Dillon and of the Company of or incidental to the negociation preparation and execution of the recited Agreement for a Lease the Building Agreement and this Agreement.

7 The Company shall in respect of the shares to be allotted to the said Arthur Dillon cause this Agreement or some other sufficient contract or Contracts within the meaning of section 7 of the Companies Act 1900 and also the returns required by that section to be filed with the Registrar of Joint Stock Companies.

8 In the event of this agreement not being confirmed and ratified by the Company on or before the day of next the said Arthur John McMillan or the said Arthur Dillon may at any time thereafter by notice in writing to be sent to the other of them determine this agreement and thereupon this Agreement shall be of no effect and neither party hereto shall have any claim for damages or otherwise howsoever in consequence of the non-performance thereof

In witness whereof the said parties have hereunto set their hands

Witness to the signature of Arthur Dillon
R A Draper,
17 Tower Royal,
Cannon Street E.C.
Clerk to Messrs Cox Laford Solicitors

Arthur Dillon

A J McMillan

Witness to the signature of Arthur John McMillan
R A Draper

Plate 6: The two-page document, dated 31st August 1905, concerning the proposed five directors as to their qualifications; it also contains the initial issue allocation of 250 shares to each of them.
Courtesy National Archives; ref – BT 31/17576/86120.

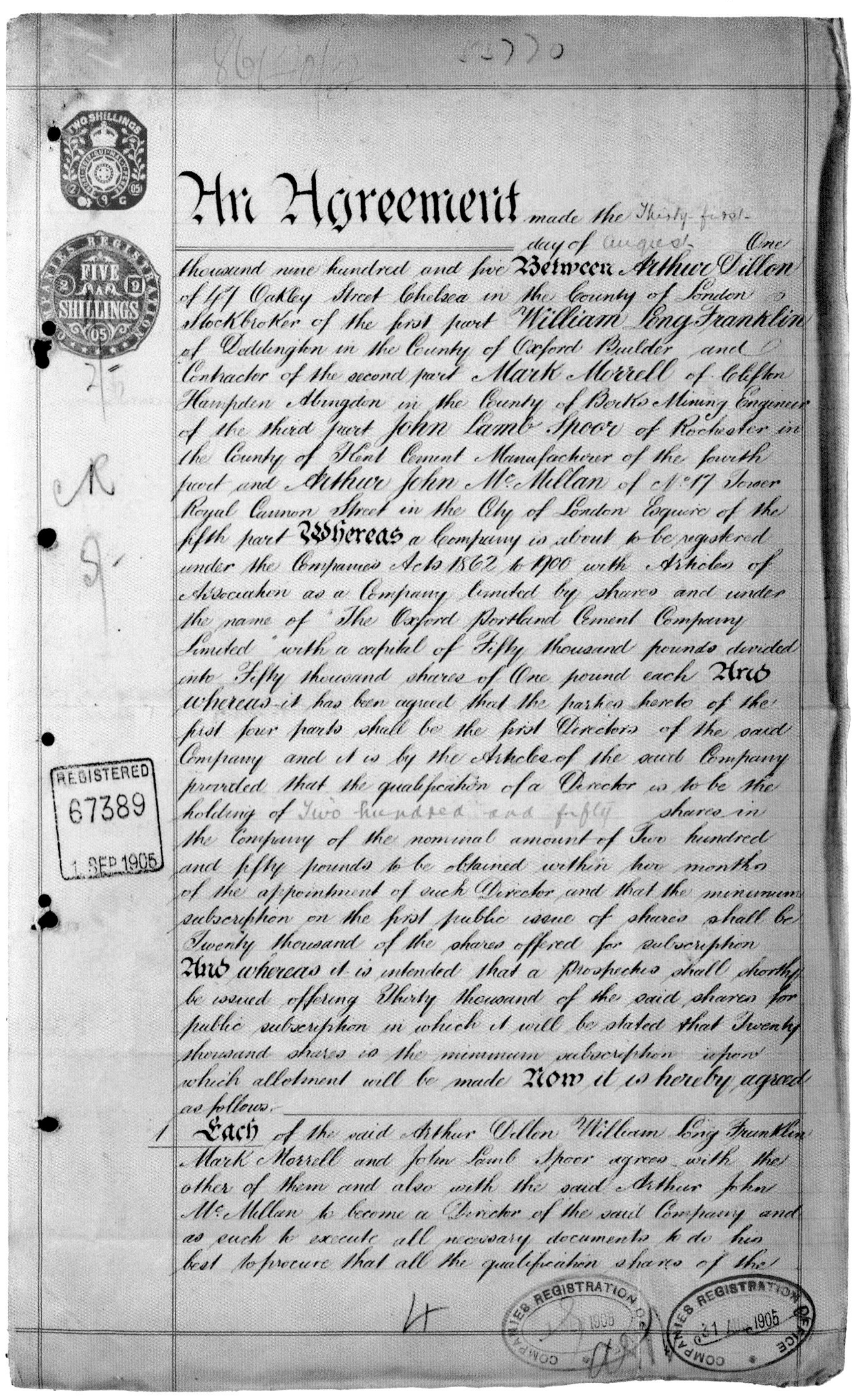

An Agreement made the Thirty-first day of August One thousand nine hundred and five Between Arthur Dillon of 47 Oakley Street Chelsea in the County of London a Stockbroker of the first part William Long Franklin of Deddington in the County of Oxford Builder and Contractor of the second part Mark Morrell of Clifton Hampden Abingdon in the County of Berks Mining Engineer of the third part John Lamb Spoor of Rochester in the County of Kent Cement Manufacturer of the fourth part and Arthur John McMillan of No 17 Tower Royal Cannon Street in the City of London Esquire of the fifth part Whereas a Company is about to be registered under the Companies Acts 1862 to 1900 with Articles of Association as a Company limited by shares and under the name of 'The Oxford Portland Cement Company Limited' with a capital of Fifty thousand pounds divided into Fifty thousand shares of One pound each And whereas it has been agreed that the parties hereto of the first four parts shall be the first Directors of the said Company and it is by the Articles of the said Company provided that the qualification of a Director is to be the holding of Two hundred and fifty shares in the Company of the nominal amount of Two hundred and fifty pounds to be obtained within two months of the appointment of such Director and that the minimum subscription on the first public issue of shares shall be Twenty thousand of the shares offered for subscription And whereas it is intended that a Prospectus shall shortly be issued offering Thirty thousand of the said shares for public subscription in which it will be stated that Twenty thousand shares is the minimum subscription upon which allotment will be made Now it is hereby agreed as follows:-

1 Each of the said Arthur Dillon William Long Franklin Mark Morrell and John Lamb Spoor agrees with the other of them and also with the said Arthur John McMillan to become a Director of the said Company and as such to execute all necessary documents to do his best to procure that all the qualification shares of the

REGISTERED 67389 1 SEP 1905

COMPANIES REGISTRATION OFFICE 31 AUG 1905

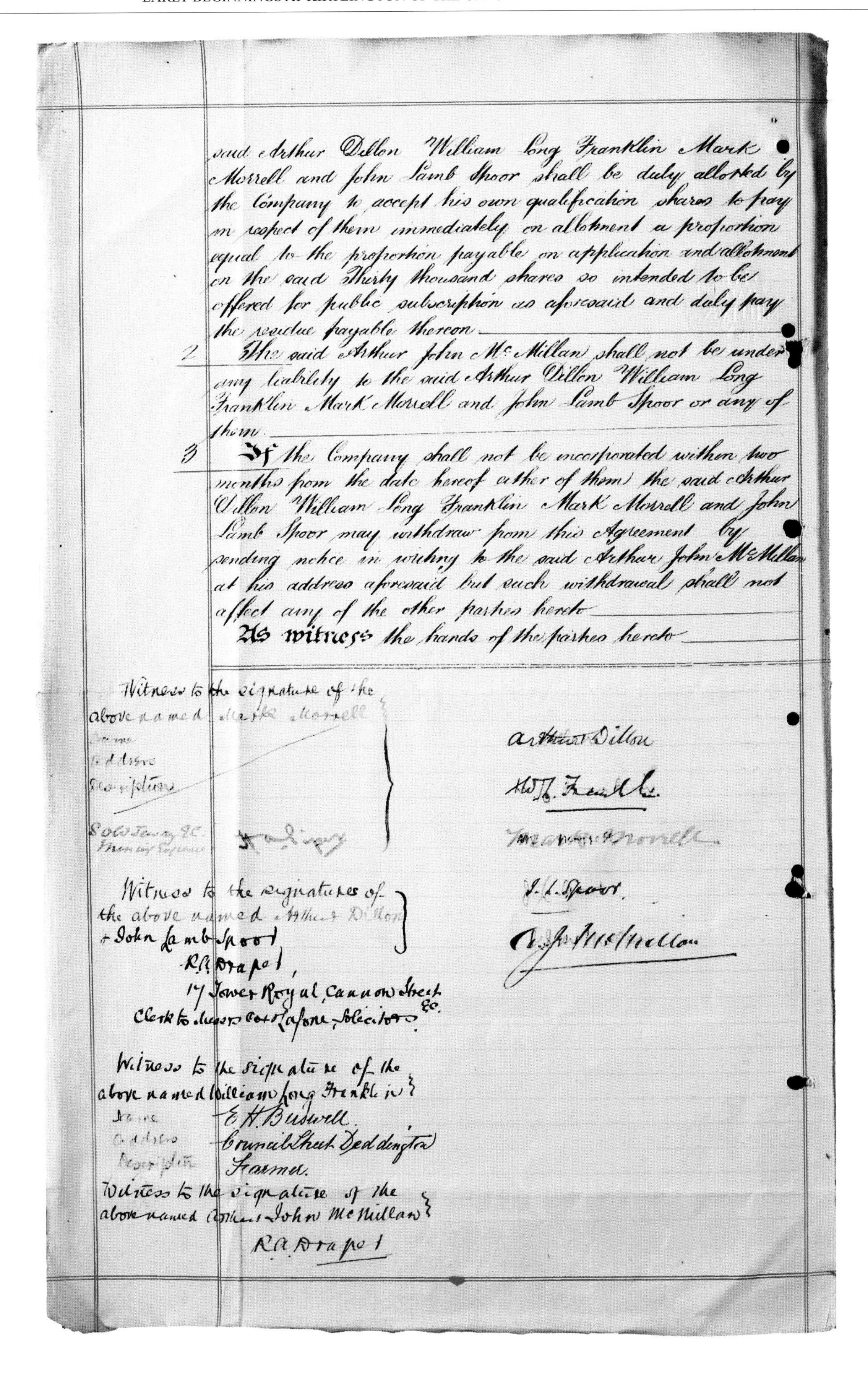
said Arthur Dillon William Long Franklin Mark Morrell and John Lamb Spoor shall be duly allotted by the Company to accept his own qualification shares to pay in respect of them immediately on allotment a proportion equal to the proportion payable on application and allotment on the said Thirty thousand shares so intended to be offered for public subscription as aforesaid and duly pay the residue payable thereon

2 The said Arthur John McMillan shall not be under any liability to the said Arthur Dillon William Long Franklin Mark Morrell and John Lamb Spoor or any of them

3 If the Company shall not be incorporated within two months from the date hereof either of them the said Arthur Dillon William Long Franklin Mark Morrell and John Lamb Spoor may withdraw from this Agreement by sending notice in writing to the said Arthur John McMillan at his address aforesaid but such withdrawal shall not affect any of the other parties hereto

As witness the hands of the parties hereto

Witness to the signature of the above named Mark Morrell
Name
Address
Description

Witness to the signatures of the above named Arthur Dillon & John Lamb Spoor
R. A. Draper,
17 Tower Royal, Cannon Street, EC.
Clerk to Messrs Cox & Lafone, Solicitors.

Witness to the signature of the above named William Long Franklin
Name E. H. Buswell.
Address Council Street Deddington
Description Farmer.

Witness to the signature of the above named Arthur John McMillan
R. A. Draper

Arthur Dillon
W. L. Franklin
Mark Morrell
J. L. Spoor.
A. J. McMillan

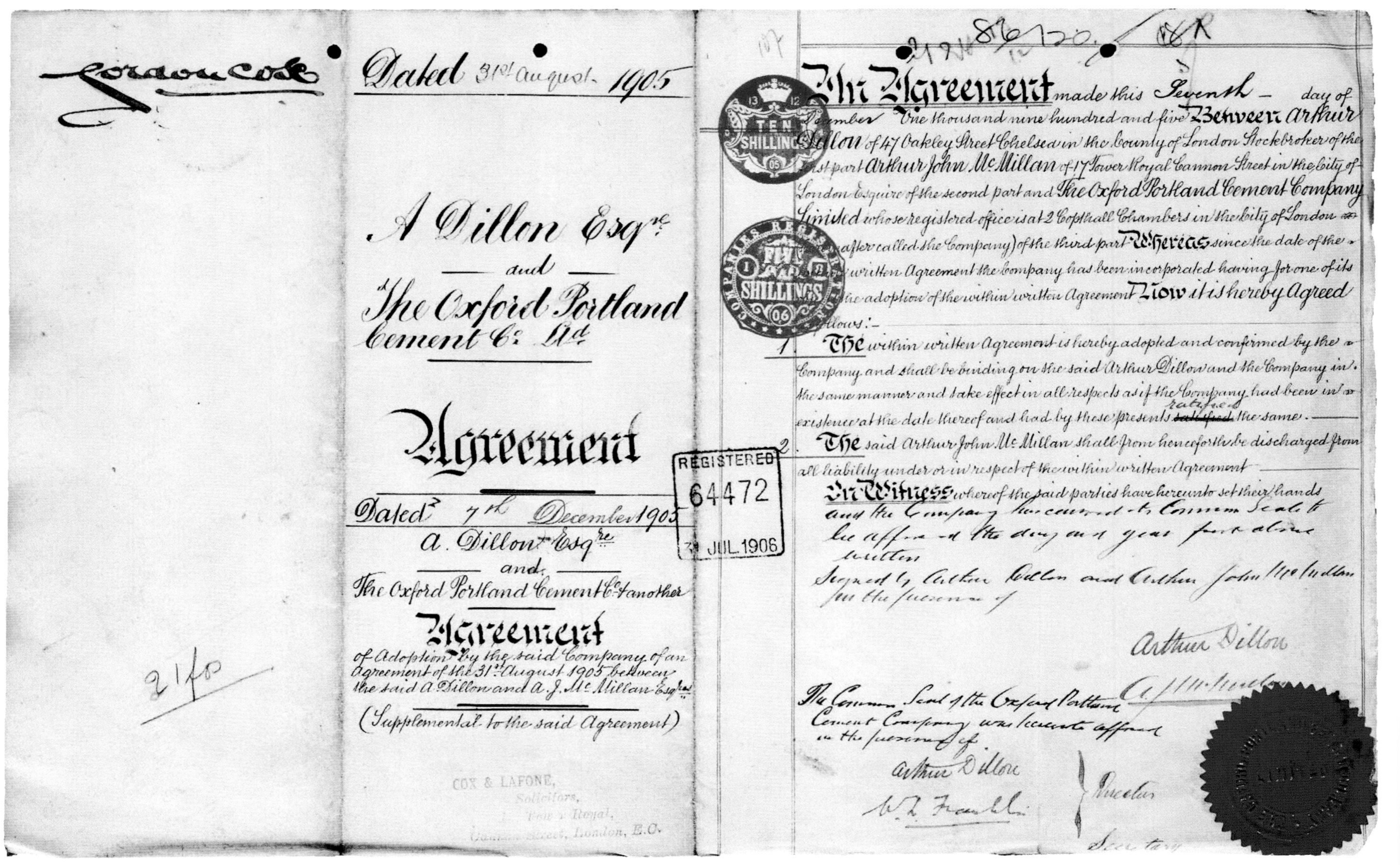

Gordon Cole

Dated 31st August 1905

A Dillon Esq

and

The Oxford Portland Cement Co Ltd

Agreement

Dated 7th December 1905

A. Dillon Esq &c

and

The Oxford Portland Cement Co & another

Agreement

of Adoption by the said Company of an Agreement of the 31st August 1905 between the said A Dillon and A. J. McMillan Esq

(Supplemental to the said Agreement)

COX & LAFONE, Solicitors, [illegible] Royal, [illegible] Street, London, E.C.

REGISTERED 64472 JUL. 1906

This Agreement made this Seventh day of December One thousand nine hundred and five Between Arthur Dillon of 47 Oakley Street Chelsea in the County of London Stockbroker of the first part Arthur John McMillan of 17 Tower Royal Cannon Street in the City of London Esquire of the second part and The Oxford Portland Cement Company Limited whose registered office is at 2 Copthall Chambers in the City of London (hereinafter called the Company) of the third part Whereas since the date of the within written Agreement the Company has been incorporated having for one of its objects the adoption of the within written Agreement Now it is hereby Agreed as follows:-

1. The within written Agreement is hereby adopted and confirmed by the Company and shall be binding on the said Arthur Dillon and the Company in the same manner and take effect in all respects as if the Company had been in existence at the date thereof and had by these presents ratified the same.

2. The said Arthur John McMillan shall from henceforth be discharged from all liability under or in respect of the within written Agreement

In Witness whereof the said parties have hereunto set their hands and the Company has caused its Common Seal to be affixed the day and year first above written

Signed by Arthur Dillon and Arthur John McMillan in the presence of

Arthur Dillon

A J McMillan

The Common Seal of the Oxford Portland Cement Company was hereunto affixed in the presence of

Arthur Dillon } Director

W. L. Frankl[illegible] Secretary

Plate 7: This further agreement dated 7th of December 1905 in which A. Dillon and Arthur John McMillan seemed to have had a falling-out, as it states that McMillan 'shall from henceforth be discharged from all liability under or in respect written agreement.' He does not appear as a director at any time after this agreement. Courtesy National Archives; ref – BT 31/17576/86120.

The actual registration of the Oxford Portland Cement Company No. 86120 took effect from 12th of October 1905, when it was incorporated under the Companies Act 1900 – *see Plate 8*.

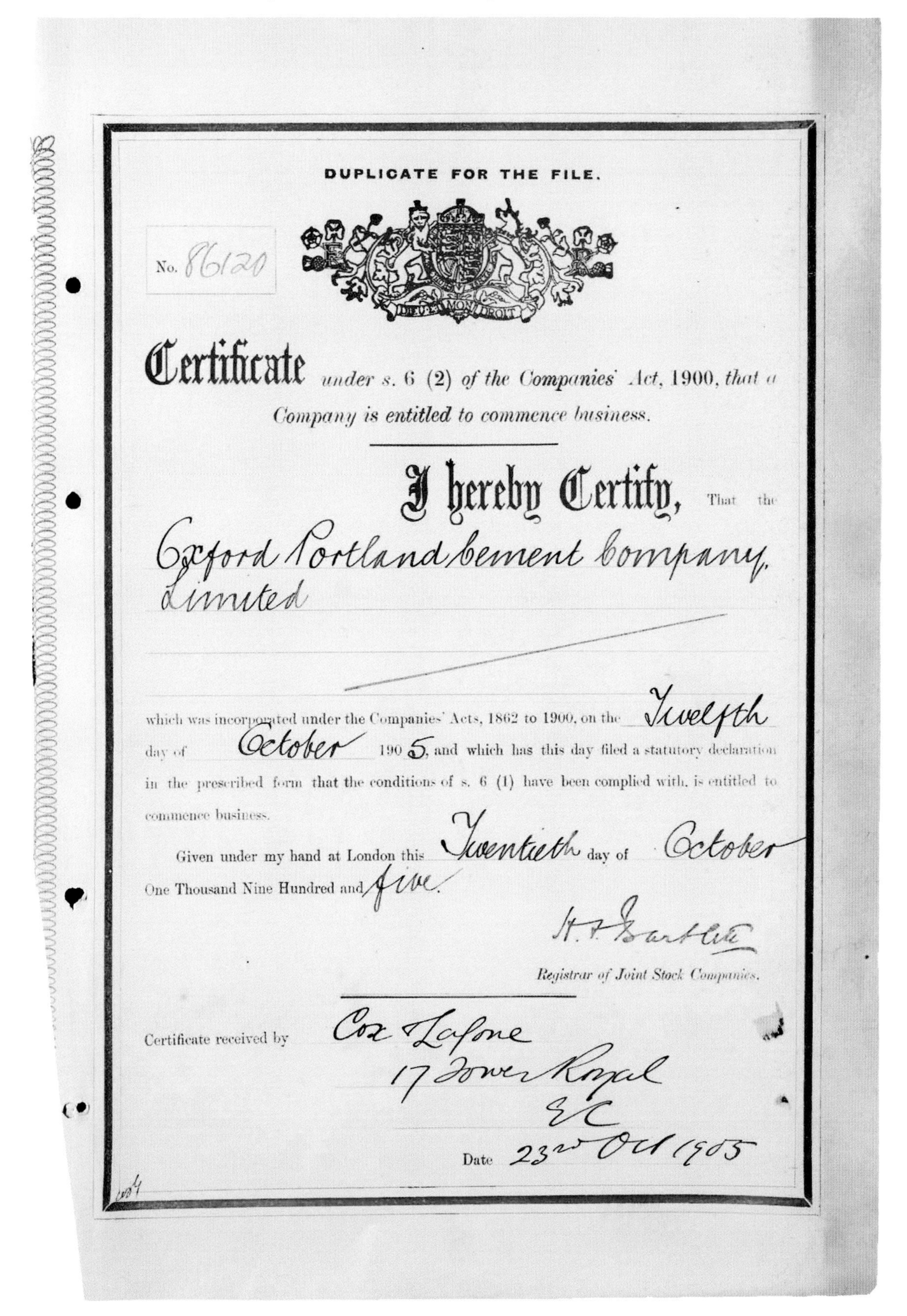
DUPLICATE FOR THE FILE.

No. 86120

DIEU ET MON DROIT

Certificate under s. 6 (2) of the Companies' Act, 1900, that a Company is entitled to commence business.

I hereby Certify, That the Oxford Portland Cement Company, Limited

which was incorporated under the Companies' Acts, 1862 to 1900, on the Twelfth day of October 1905, and which has this day filed a statutory declaration in the prescribed form that the conditions of s. 6 (1) have been complied with, is entitled to commence business.

Given under my hand at London this Twentieth day of October One Thousand Nine Hundred and five.

H. F. Bartlett

Registrar of Joint Stock Companies.

Certificate received by Cox & Lafone 17 Tower Royal EC

Date 23rd Oct 1905

Plate 8: Facsimile reproduction of the Certificate for the Company 86120 – Registration of the OPCC Ltd. on the 12th of October 1905. Courtesy National Archives; ref – BT 31/17576/86120.

Plate 9: The complete prospectus for the Oxford Portland Cement Company Limited and shares available. It contains all the relevant details of the land and rents due and makes very interesting reading.

Courtesy National Archives; ref – BT 31/17576/86120.

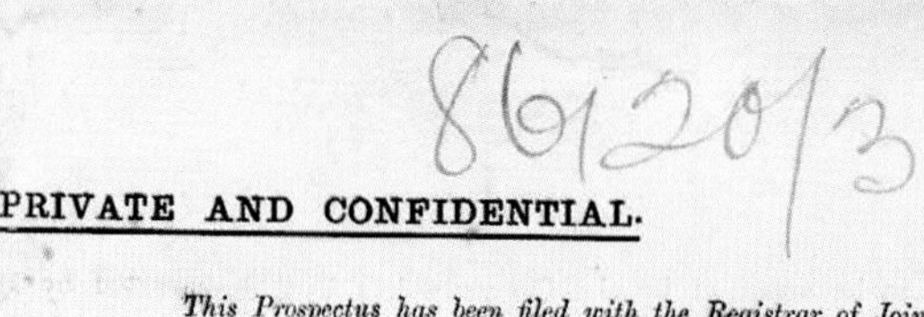

PRIVATE AND CONFIDENTIAL.

This Prospectus has been filed with the Registrar of Joint Stock Companies.

THE OXFORD PORTLAND CEMENT COMPANY, LIMITED

(To be Incorporated under the Companies Acts, 1862 to 1900).

CAPITAL - - £50,000,

DIVIDED INTO

50,000 Shares of £1 each,

of which 30,000 Shares are now offered for Subscription at par,

PAYABLE AS FOLLOWS:—

5s. per Share on Application,
5s. " " Allotment,

and the Balance as required by Calls of not more than 5s. per Share, at intervals of not less than Two Months.

Shareholders may pay up in full on Allotment, and on the amount paid up in advance of Calls interest will be allowed at the rate of 5 per cent. per annum.

2,000 Shares will be issued to the Vendor as fully paid as hereinafter mentioned, and 18,000 will be held in reserve for future issue.

REGISTERED 67387 1 SEP 1905

The minimum Subscription on which the Directors may proceed to Allotment is £20,000.

Directors.

ARTHUR DILLON, Esq., 47, Oakley Street, Chelsea, S.W., Stockbroker.
WILLIAM LONG FRANKLIN, Esq., Deddington, Oxon, Builder and Contractor.
MARK MORRELL, Esq., Clifton Hampden, Abingdon, late of Buluwayo, Mining Engineer.
*JOHN LAMB SPOOR, Esq., J.P., Rochester, Cement Manufacturer.

*Will join the Board on the completion of the works to be executed under the Contract (No. 2) hereinafter referred to.

Bankers.

MESSRS. BARCLAY & COMPANY, LIMITED (Old Bank), Oxford.

Solicitors.

MESSRS. COX & LAFONE, Tower Royal, Cannon Street, London, E.C.

Auditor.

H. W. HOLLAND, Esq., Chartered Accountant, 60, Haymarket, London, W.

Secretary and Offices *(pro tem.)*.

R. G. REYNOLDS, 2, Copthall Chambers, London, E.C.

PROSPECTUS.

THIS Company is intended to be formed for the purpose, amongst other things, of acquiring a Lease of certain Land known as the "Washford Pits," and rights over adjoining Limestone and Clay-bearing Lands at Kirtlington, near Oxford, and of erecting Works, and manufacturing Portland Cement thereon.

The property proposed to be acquired by the Company has been inspected by Mr. Bertram Blount, F.I.C., of Westminster, who has carefully analysed the Limestone and Clay found thereon, and certifies that these materials are of a suitable quality for the manufacture of Portland Cement of good quality.

Mr. J. L. Spoor, J.P., Portland Cement Manufacturer, of Rochester, has also inspected the property, and reports that the raw materials exist in ample and practically inexhaustible quantities, and are suitable for the manufacture of first-class Portland Cement at a cost which will compare favourably with that ruling on the Thames and Medway.

The site of the Works immediately adjoins the Oxford Canal, which affords cheap access to large and important centres of consumption, and secures a ready market.

In the first instance it is proposed to provide for an output of 10,000 tons of Cement per year; but in designing the Works the probable necessity of increasing the output later on will be kept in view, as it is believed there is ample scope for a large extension in the districts to be served.

In a large area, including many important markets, the freight on Cement from the Company's Works will compare favourably with that payable from the nearest competing Works.

The Land forms part of the Settled Estates of Sir George James Egerton Dashwood, Bart., of Kirtlington Park, Oxford, who has agreed to grant the necessary Lease of the Land for the Works, and of the Limestone and Clay-getting rights on advantageous terms to the Company. The Land for the Works will comprise about five acres, and will be leased for a term of 60 years, at a rent of £2 per acre, with the exception of the Land to be occupied by Workmen's Dwelling Houses, for which the rent is to be £5 per acre. The Limestone and Clay-getting rights will extend over a large portion of adjoining Land, where valuable deposits of Limestone and Clay exist, and will be leased for the same term of 60 years. The Royalties on Cement manufactured at the Works are fixed at 1s. per ton, and the Royalties on Limestone or Clay used for any other purpose than Cement-making are fixed at 6d. per ton. The minimum Royalties are fixed at £200 for the first year, and £300 for each succeeding year. On the completion of the Works it is fully believed the Company will be in a position to earn at least 10 per cent. on the amount of its issued capital, even when prices rule low, and that in good times this return will be considerably exceeded. The cost of raw material is fixed under the above-mentioned Lease, and the only fluctuating items in cost of manufacture are coal and labour, and this within narrow limits.

As set out in the Contract mentioned below, Mr. Spoor has agreed to design and erect the Works on the most modern and economical lines and to hand them over as a going concern, and will for the period of at least three years give the Company the benefit of his experience and advice in the manufacture and organization of the business. Mr. Spoor has also agreed to join the Board of Directors.

In addition to the manufacture of Cement, the Company proposes to quarry for building purposes the Limestone Rock found on the land. These quarries are in use at the present time, and are stated to have been continuously and profitably worked for upwards of a century. This should materially assist the dividend during the first year when the Works are in course of erection.

2,000 Shares will be issued as fully paid to Mr. Arthur Dillon, of 47, Oakley Street, Chelsea, Stockbroker, the Vendor to the Company, in consideration of his services in securing the above-mentioned Lease and negotiating the Contract with Mr. Spoor, hereinafter mentioned.

The contents of the proposed Memorandum of Association, the names and addresses and description of the signatories, and the number of shares subscribed for by them respectively will be found on the last page of the Prospectus.

The Articles of Association of the Company provide that the qualification of a Director is to be the holding of Shares in the Company to the nominal value of not less than £250; that the remuneration of the Directors, other than the Managing Director, shall be at the rate of £500 per annum, to be divided among them as they shall agree; that any Director being called upon to perform extra services may be paid special remuneration for such services as the Board may think fit; that the Directors may appoint a Managing Director and fix his remuneration; and that if the Company shall offer any of its Shares to the public for subscription, the Directors may exercise the powers conferred by Section 8 of the Companies Act, 1900, but so that the commission shall not exceed 5 per cent. on the Shares in each case offered.

The preliminary expenses in connection with the formation of the Company, including fees payable on registration, the *ad valorem* stamp duty, law costs, printing, &c., will be paid by the Company and are estimated at £300.

The following Contracts have been entered into:—(1) Dated the 31st day of August, 1905, and made between Sir George James Egerton Dashwood, Bart., of the one part, and Arthur Dillon of the other part, providing for the granting of the Lease before-mentioned; (2) Dated the 2nd day of August, 1905, and made between John Lamb Spoor of the one part, and Arthur Dillon of the other part, providing for the erection and complete equipment of the Works for an agreed sum of £20,000; (3) Dated the 31st — day of August, 1905, and made between Arthur Dillon of the one part, and Arthur John McMillan for and on behalf of the Company of the other part, providing for the transfer from Mr. Dillon to the Company of all his rights under the two above-mentioned Contracts, in consideration of the issue to him of 2,000 fully-paid Shares of the Company; (4) Dated the 31st day of August, 1905, and made between Arthur Dillon of the first part, William Long Franklin of the second part, Mark Morrell of the third part, John Lamb Spoor of the fourth part, and Arthur John McMillan of the fifth part, providing for the acquisition by the parties of the first four parts of their qualification shares.

WLF
MM
AD
JLS

The above-mentioned Contracts and Reports and the Company's proposed Memorandum and Articles of Association can be seen at the offices of the Solicitors at any time during business hours.

Application for shares should be made on the form accompanying this Prospectus and sent with a cheque for the deposit of 5/- per Share to the Company's Bankers or Secretary. If the whole amount of shares applied for be not allotted, the surplus paid on application will be applied towards the amount due on allotment, and any balance will be returned.

DATED, 2nd September ~~August,~~ 1905.

Arthur Dillon

William Long Franklin

Mark Morrell

J. L. Spoor.

NAMES, ADDRESSES AND DESCRIPTIONS OF ALLOTTEES.

Surname.	Christian Name.	Address.	Description.	Number of Shares allotted. Preference.	Ordinary.	Deferred.	Founders.
Brisco	Hylton Ralph	The Cottage, Foston, Derby	J.P.	500			
Brisco	Lilian Mabel	do	Married Woman	10			
Cheeseman	Violet	The Mount, Claydon, Surrey	Wife of W.E.F. Cheeseman	10			
Brisco	Hilda Cunningham	Crofton Hall, Wigton, Cumberland	Spinster	10			
Browne	Cyril Edward	Gardenhill, Castleconnell, Limerick	Government Auditor	50			
Selby-Bigge	Edith Lindsay	7, Wilbraham Place, London, S.W.	Married Woman	100			
Dillon	Hilda E.	47, Oakley St. Chelsea	Spinster	100			
Lucas-Tooth	Robert Lucas	1, Queen's Gate, S.W.	Gentleman	5000			
Dashwood	George Lionel	1, Fleet Street, E.C.	Banker	100			
Bury	Ralph Frederic	St Leonards, Nazeing, Essex	Esquire	200			
Bury	Anna Loftus	do	Widow	200			
Blount	Oscar	The Hut, Windsor	Esquire	3000			
Browne	Anthony Howe	Manor House, Sutton, Middlesex	a Minor (son of Percy H. Browne)	10			
Browne	Margaret Kathleen	do	a Minor	20			
Wyatt	William West	67, St Giles's, Oxford	Builder	100			
Compton	E D	Summer Fields, St Leonards-on-Sea	Schoolmaster	50			
Dillon	Margaret Sophia	47, Oakley St Chelsea, S.W.	Spinster	40			
Dillon	Hilda	do	do	15			
Scott	John Oliver	Kirtlington, Oxford	Land Agent	25			
Richards	Alice	Westridge, Ryde, I.W.	Married Woman	1000			
Page	William	Mounds House, Deddington, Oxford	Farmer	50			
Dashwood	Sir George J. E.	Kirtlington Park, Oxford	Bart.	5000			
Franklin	William Long	Deddington, Oxon	Builder	250			
Benfield	George	Iffley, nr Oxford	Builder & Contractor	250			
Saxley	William Frederick	65, Divinity Road, Oxford	do	250			
Spoor	John Lamb	Rede Court, Rochester	Cement Manufacturer	3000			
Dillon	Arthur	2, Copthall Chambers, E.C.	Stockbroker	260			
Bazley	Gardner Sebastian	Hatherop Castle, Fairford, Glos.	J.P.	200			
Morrell	Mark	Clifton Hampden, nr Abingdon	Gentleman	240			
Hands	Arthur Edward	New St. Deddington	Esquire	200			
				20,240			

This margin is reserved for binding, and should not be written across.

This margin is reserved for binding, and should not be written across.

Signature Edward Powell

Description Secretary

To the Registrar of Joint Stock Companies.

(a) Here insert "I" or "We."

(a) We, the undersigned, hereby give you notice, pursuant to s. 2 (2) of the Companies' Act, 1900, that the following persons have consented to be Directors of the Oxford Portland Cement Company Limited.

Name	Address	Description
Arthur Dillon	47 Oakley St Chelsea SW	Stockbroker
William Long Franklin	Deddington, Oxon	Builder & Contractor
Mark Morrell	Clifton Hampden Nr Abingdon	Mining Engineer
John Lamb Spoor	Rochester	Cement Manufacturer

Plate 10: The document above lists the first £1 shareholders and it is interesting to note just how many 'Dillons' appear, plus Sir George John Dashwood's huge investment of £5000.
Courtesy National Archives; ref – BT 31/17576/86120.

Plate 11: The document (left) lists officially the four directors of the Oxford Portland Cement Company Limited in October 1905, with the name of Arthur John McMillan missing.
Courtesy National Archives; ref – BT 31/17576/86120.

Plate 12: A photograph commemorating the 100th birthday of I.C. Johnson but showing OPCC director and contractor John Lamb Spoor in position third from the left – back row. *Courtesy Chris Down collection.*

Deputation of Cement Manufacturers to I. C. Johnson on the occasion of his 100th Birthday, 28th January, 1911.

Standing: F. ANTHONY WHITE; E. CHARLETON; J. L. SPOOR; HUDSON EARLE.
Sitting: A. TOLHURST; C. H. WATSON; I. C. JOHNSON; C. CHARLETON; J. BAZLEY WHITE.

Footnote: John Lamb Spoor (1856-1918).
Spoor was an autodidact of lowly origins whose potential was recognised by Isaac Charles Johnson – a man of similar background. Spoor was born in Swalwell, County Durham, a few miles west of Gateshead. His immediate relatives were blacksmiths associated with the local iron works. He entered employment with I.C. Johnson & Co around 1873 and moved to Cliffe, Kent in 1877 to supervise the commissioning of the cement chamber kilns there and at Greenhithe. Around 1889 he left Johnsons and set up on his own, constructing Borstal Manor works, but unfortunately running out of money before the plant was complete. His mortgagees foreclosed and he was declared bankrupt on the 26th October 1894. He then went to India and supervised the building of cement works at Madras and Bangalore. He made sufficient money after four years to return to England in 1898 and pay off his creditors in full. Returning to India he erected his own works in Calcutta but then came back to England in 1902 to set up as a cement works consultant; supervising, among other things, the building of cement works at Cousland and KIRTLINGTON in Oxfordshire at the ripe old age of 50, plus the rebuilding of the Lyme Regis cement works, while living in Strood, Kent. He died in 1918 and was buried in Cliffe churchyard.
The obituary of Mr J.L. Spoor, Rede Court, Stroud, Kent was recorded in the Maidstone Telegraph in 1918 and briefly stated that he was a Military representative of the Chatham Recruiting Area. His great attribution was the organisation of the Kent Prisoners of War Fund supervising the despatch of food parcels, clothing etc to the men of Kent who were captives in enemy countries. In business he was for a greater part of his life, engaged in the CEMENT INDUSTRY.

The Dashwood family left Kirtlington Park Estate during 1909 and the estate was purchased by the Earl of Leven and Melville, who again in 1922 sold it onto Herbert Maitland Budgett. The land and farms appear not to have been included in this sale of 1922 and so the solicitors tried to tidy up leases over the property and lease to the Kirtlington/ Washford Pits was sold outright (through an agent John White) to the Oxford Portland Cement Company Limited.

An extract from the 1922 sale records, listed the annual rent to OPCC had been £358 16s 8d, plus a further £1 10s to the solicitors for handling this rent. There was also a mineral extraction rent averaging about £310 during the period 1914-1922.

On a local note; the magnificent Oxford Christ Church Cathedral and College were constructed in the mid-16th century with material that had been supplied from a small lime kiln near to Kirtlington village, showing the historical value of the local mineral beds in the area.

In April 1910 the OPCC advertised in the Oxford Times local newspaper: – 'The Oxford Portland Cement Co., Ltd., Beg to announce that they are now selling their best PORTLAND CEMENT at 3/- per bag, cash, at their Warehouse in WORCESTER STREET, besides the usual one-ton lots'.

The Oxford Portland
Cement Co., Ltd.,

Beg to announce that they are
now selling their best

PORTLAND CEMENT at 3/- per bag, cash,

at their Warehouse in WORCESTER STREET,
besides the usual one-ton lots.

In June 1911, the Oxford Portland Cement Co. Ltd. became better known locally, after having exhibited on stand No. 21 at the Oxford Agricultural Show, held in Kidlington, Oxfordshire. Their headboard on the stand read with similar wording to their previous advert: 'The Oxford Portland Cement Co., Ltd., are selling their celebrated Oxford Brand of Portland Cement at their warehouse beside the canal basin wharf in Worcester Street, Oxford.'

The construction of the Oxford Portland Cement Co. Ltd. works at Kirtlington (sometimes referred to as the Washford Pits Works) commenced in late 1905, finally coming on-line in 1907.

Plate 13: The Oxford Portland Cement Works at Kirtlington (Washford Pits) under construction in 1906 with the Oxford Canal evident on the left, winding away into the distance towards Banbury. The picture has a note that Lt. Col. H.M. Shellard was the photographer.
Courtesy Oxfordshire County Council, Oxford History Centre; ref POX0208526.

The initial registered company address was at 2 Capthall Chambers, London, then in August 1907 it changed to 27 Cannon Street, London EC but this again was changed finally in March 1909, to the Kirtlington Works site. This was at a critical time when the Oxford Canal was being upgraded and deepened to being able to take larger and deeper canal barges all the way south to Oxford and north to Coventry and Birmingham.

The OPCC works during the remaining part of 1907 was able to produce 481 tons of portland cement, which was hand packed and sold in returnable one cwt hessian sacks.

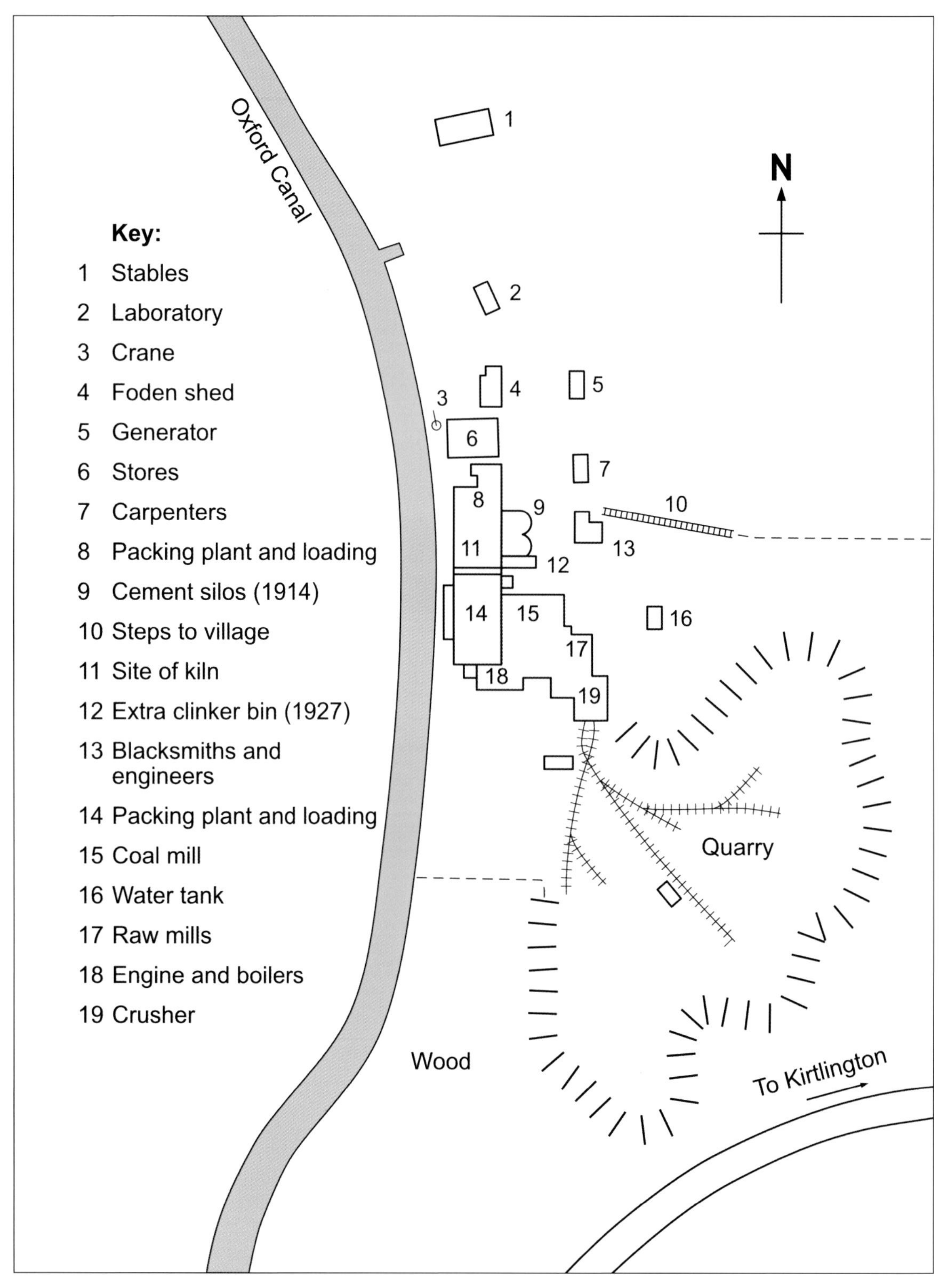

Figure 3: A diagrammatic representation plan of the Kirtlington/Washford Pits cement works based on information from contemporary records known and OS maps of the area.

One of the first batches of cement produced at Kirtlington was used in the construction of a small concrete building on A.H. Dillon's land and bore the company's trade mark, which consisted of an Ox standing above water, *(see trademark on the first page of the book).* Funnily enough, this was very similar to another famous Oxford industry, the car manufacturing company; Morris Motors Ltd.

The trademark image, which changed slightly over the years, has the meaning of the ancient form of Oxford – OXENFORD.

The next year saw a vast improvement in the works and the way the cement was made.

The increase in output prompted the company in April 1909 to establish a sales depot in Oxford and it asked the Oxford Canal Company permission to build a warehouse at their Worcester Street wharf. By September of that year a building was constructed against the Worcester Street Bridge *(Plate 14)* at a cost of £223. Much later, in 1925, the building was equipped with an electrical hoist and a runway system to speed up the handling of the cement bags. However, there were not enough sales outlets to meet the demand for the company's products and further sales depots were added at the Coventry Canal Wharf and at the Crescent Wharf on the Newell arm of the Birmingham Canal.

Plate 14: The large building (centre) seen constructed right up against the Worcester Street hump-backed canal bridge, was built for the Oxford Portland Cement Co. Ltd. works at Kirtlington as cement-bag storage and was the company's local sales point on the Oxford Canal, at the Worcester Street wharf.

Oxfordshire County Council, Oxford History Centre ref; D268795a.

The Great Western Railway, as the forward-looking company that they were, wanted any business that the OPCC were likely to give them and so in March 1909, discussed the possibility of a transshipment facility at Bletchington* Station *(Plate 17),* initially using the Oxford Canal Wharf at Enslow, which is a hamlet next to Bletchington Station *(see Plate 16)* and was the unloading point for the Kirtlington cement works canal barges. This wharf was luckily situated right next to the Bletchington GWR main line station but not owned by the GWR so further dealings were needed to be initiated!

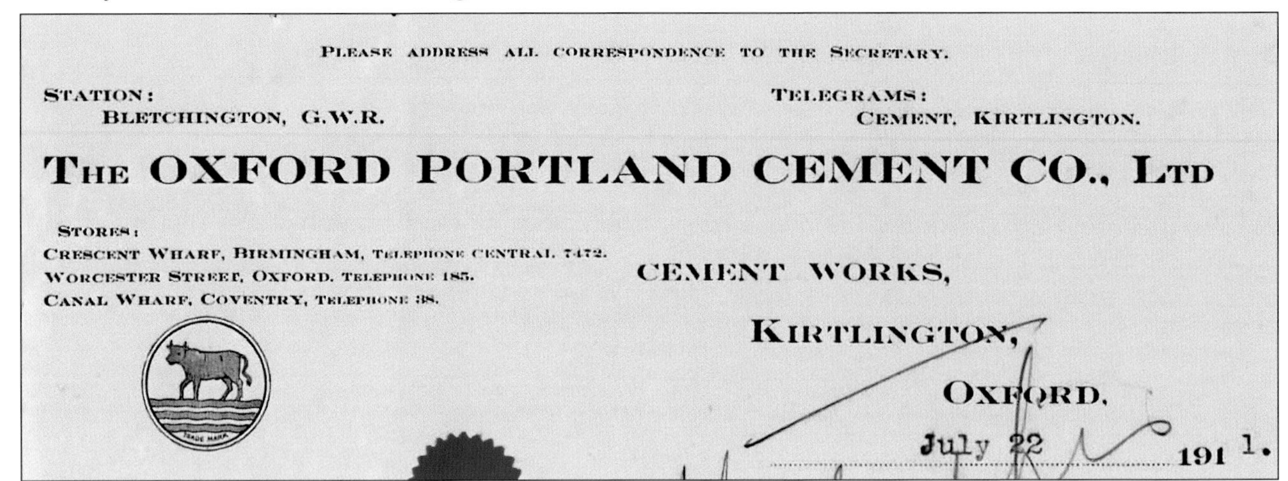

PLEASE ADDRESS ALL CORRESPONDENCE TO THE SECRETARY.

STATION:
BLETCHINGTON, G.W.R.

TELEGRAMS:
CEMENT, KIRTLINGTON.

THE OXFORD PORTLAND CEMENT CO., LTD

STORES:
CRESCENT WHARF, BIRMINGHAM, TELEPHONE CENTRAL 7472.
WORCESTER STREET, OXFORD. TELEPHONE 185.
CANAL WHARF, COVENTRY, TELEPHONE 38.

CEMENT WORKS,
KIRTLINGTON,
OXFORD,

July 22 1911.

Plate 15: An image of the Company letterhead in 1911, showing the nearest GWR Station was at Bletchington and the Company's three cement stores and sales offices at Birmingham, Oxford, and Coventry plus incorporating the Company's registered trademark. Note that there is no telephone number for the works, only telegrams! Compare with the letterhead in Plate 27 to see the progress of the company's stationery.
Courtesy National Archives; ref – BT 31/17576/86120.

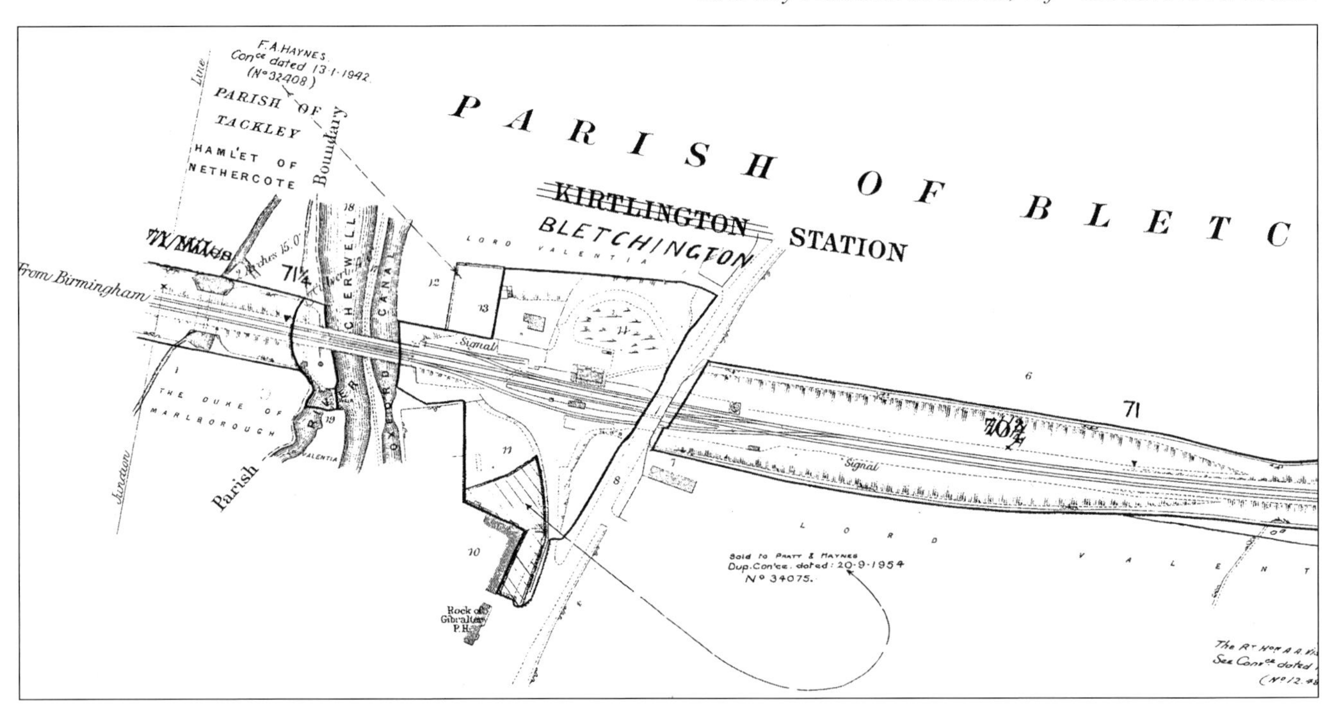

Figure 4: Part of the GWR original line survey of ownership and track layout of the Bletchington Station. Note that the survey has been altered from Kirtlington Station to Bletchington Station. *Courtesy Chris Turner collection.*

*Bletchington Station in Oxfordshire, was just over 2 km from the village of the same name. It was proposed by the Oxford to Rugby Railway but, before the station was opened in 1850, this company had been absorbed by the GWR, who built the station. The station had many names from its opening: Woodstock Road (1850), Kirtlington (1855) and finally Bletchington (1890). Note Bletchington is now named Bletchingdon and the village was and is often referred to by either name. The station was closed to passengers by British Railways in November 1964 and then June 1965 to goods traffic.

Plate 16: This is Enslow/Bletchington Canal wharf and the two-high pile of cement bags from the Kirtlington cement works can be seen alongside the lean-to building (middle left). The stone building on the far left can be seen on the six-inch OS map in Figure 9. The building in the middle background is the rear view of the famous Oxfordshire hostelry, 'The Rock of Gibraltar.'

Courtesy Oxfordshire County Council, Oxford History Centre; ref POX0072516.

In February 1910, the Oxford Canal Company agreed to sell to the GWR a small plot of land, situated on the south bank of the canal, comprising of approximately 250 square yards, for the princely sum of £40. This was to be used to construct their own wharf and the GWR then introduced this into a Bill in Parliament called the General Powers Bill, being passed through without any opposition. However, the Oxford Canal had been closed in January 1910 for a long weekend maintenance and the GWR had taken advantage of this to clear the canal bank and construct a solid, tall retaining wall at a cost to the GWR of £104.

By March 1910, the canal had been dredged so that there was a large turning circle (in canal terms, a 'winding hole') for horse-drawn barges from the Kirtlington cement plant, on their short two mile journey, to be able to turn around and return to the cement works for more cement stock possibly carrying coal back to the works. Much later some of this coal needed at the Kirtlington cement works started to come by rail into the GWR Bletchington Station and was then transferred into the canal barges to go on this short journey to the Kirtlington cement works but due to rail cost, this was very limited.

In May 1910, the GWR had prepared plans on changes to the Bletchington Station track layout, by adding two short sidings on the down side to accommodate 14 wagons, and a 200-foot platform in between these sidings that was equipped with a storage shed for the Kirtlington bagged cement. Also, on the Oxford Canal bank, a 2-ton hand crane, with a radius of 12-foot and 60ft drop was constructed to lift the cement bags from the canal barges onto the new storage platform. *(see Plates 18 and 19).*

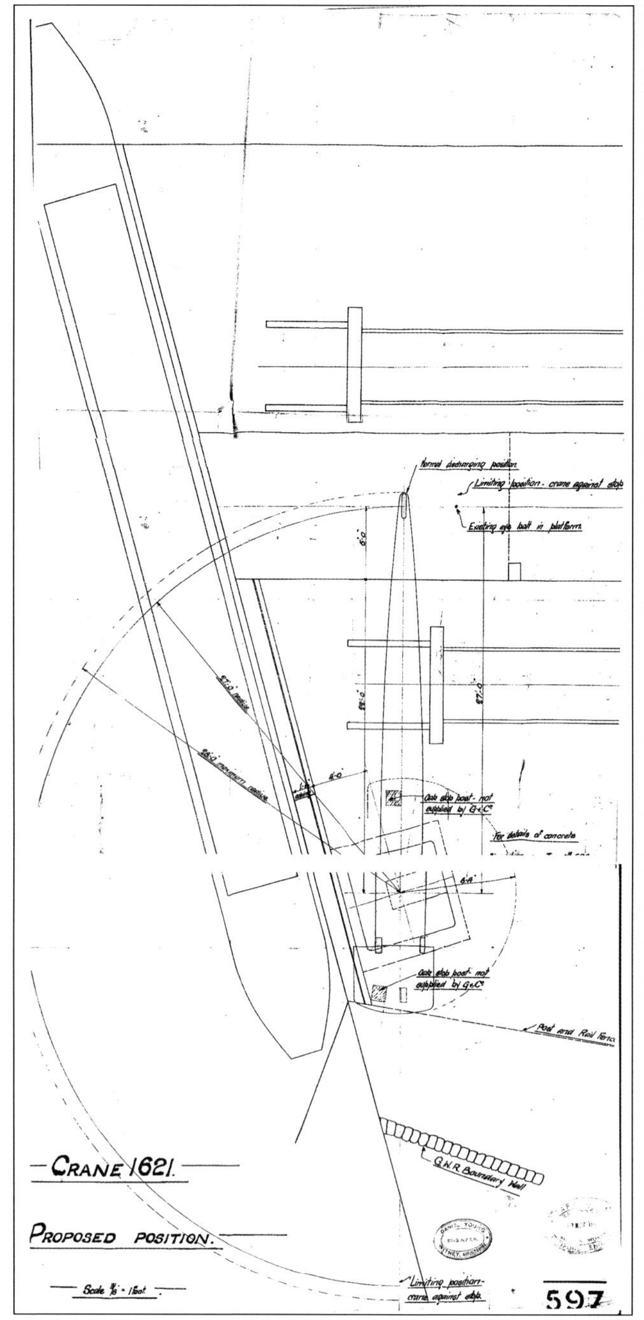

Figure 5: This is one of two Great Western Railway drawings dated 1917 showing the two-ton steam crane that was to replace the hand crane that had been installed at Bletchington Station to lift the cement bags from the OPCC canal barges from Kirtlington cement works up and onto the platform between the two new sidings. The bags were then moved by hand and housed in the cement storage transit shed, ready to be again hand loaded into trucks in the railway sidings. The Wolverhampton drawing office plan shows the position and the swinging radius of the crane and even a 'ghost' of the canal barge on the Oxford Canal alongside the canal wharf.

Courtesy Chris Down collection.

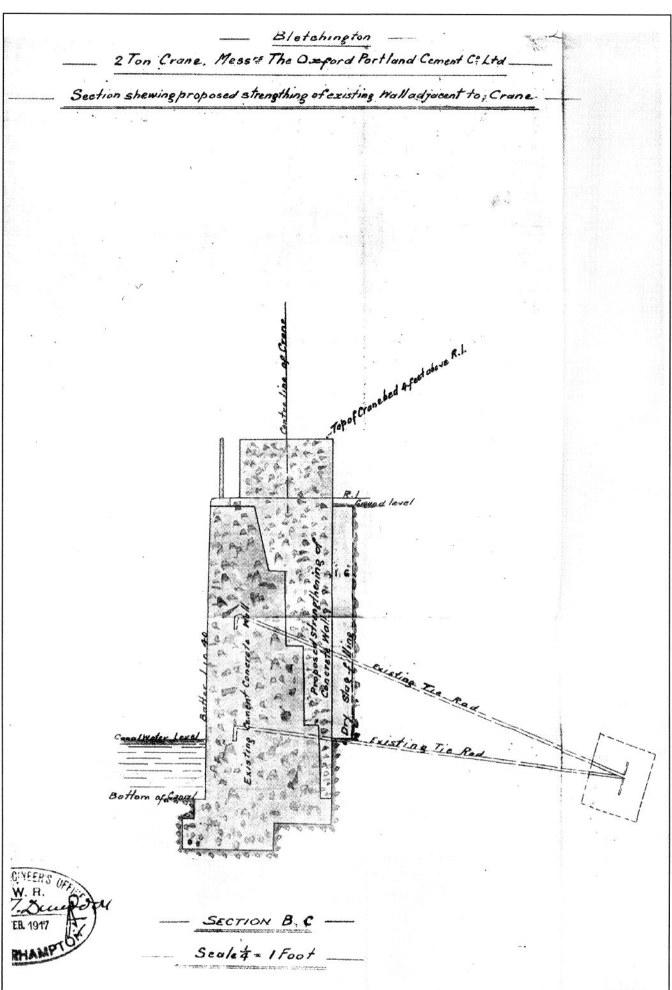

Figure 6: The second GWR plan shows how the construction of the Oxford Canal wall up to the level of the Bletchington Station is to be reinforced to take the weight of the new steam crane.

By April 1911, with a cost to the GWR of £1365, all this work was completed. The initial crane was hand operated but later in 1917, it was changed to a 2-ton steam crane, as can be seen in the three plans *Figures 5, 6 and 7.*

The evidence for this came from an obsolete deed dated 7th of December, 1917, held at Aldermarston Legal Department, BCI. *See box on page 26.*

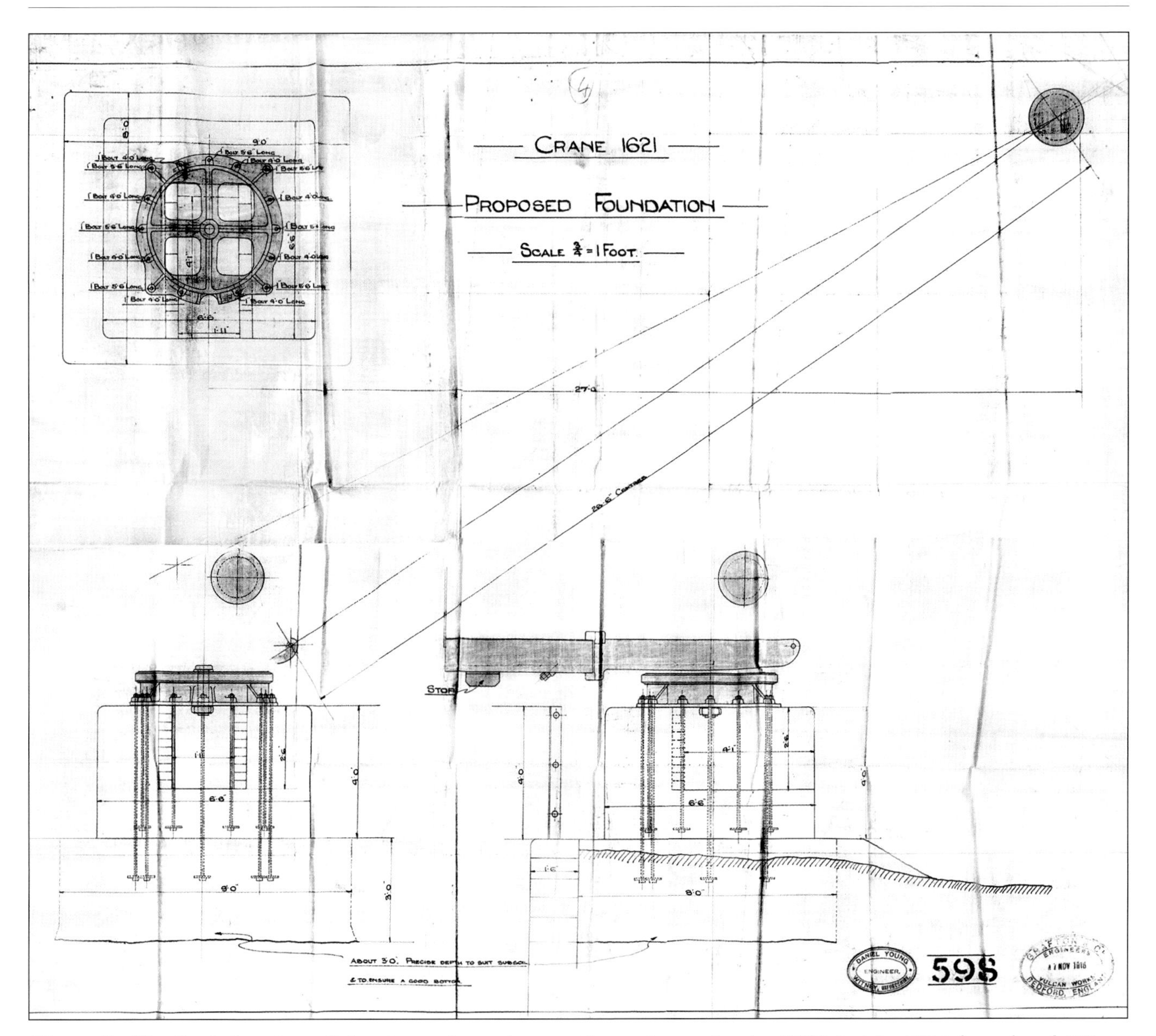

Figure 7: The final drawing of the three above, has not been drawn by the GWR but in 1916 by a local Witney Engineer Daniel Young. His company was the first engineering company to be established in Bridge Street, Witney, Oxfordshire in 1872. Daniel Young supplied and serviced steam engines and other equipment and employed three men and three boys. By 1900, he was a mechanical and consulting engineer, boiler maker, millwright and brass worker.

So, it is likely that Young acted for the OPCC as an agent and advised the OPCC that the Grafton Co. at the Vulcan works, Bedford was the best company to supply the steam crane (he probably would get a commission for his introduction). The GWR would have been involved to ensure that the necessary details were there to safeguard their property, legal right, and safety of the crane's operation. They would also need to know the lifting limits, jib radius and swing so that it would not hit anything on the tracks and Bletchington Station itself. The drawing has the stamp of Daniel Young on it before forwarding it to OPPC and this would ensure he received his commission from the crane manufacturer.

As for the year gap between the above drawing and the GWR drawings, it is not surprising, as it was during World War One plus OPCC would have had to work out what they really wanted and time to allow the scheme to sufficiently develop before speaking to the GWR to negotiate terms.

Courtesy Chris Down collection.

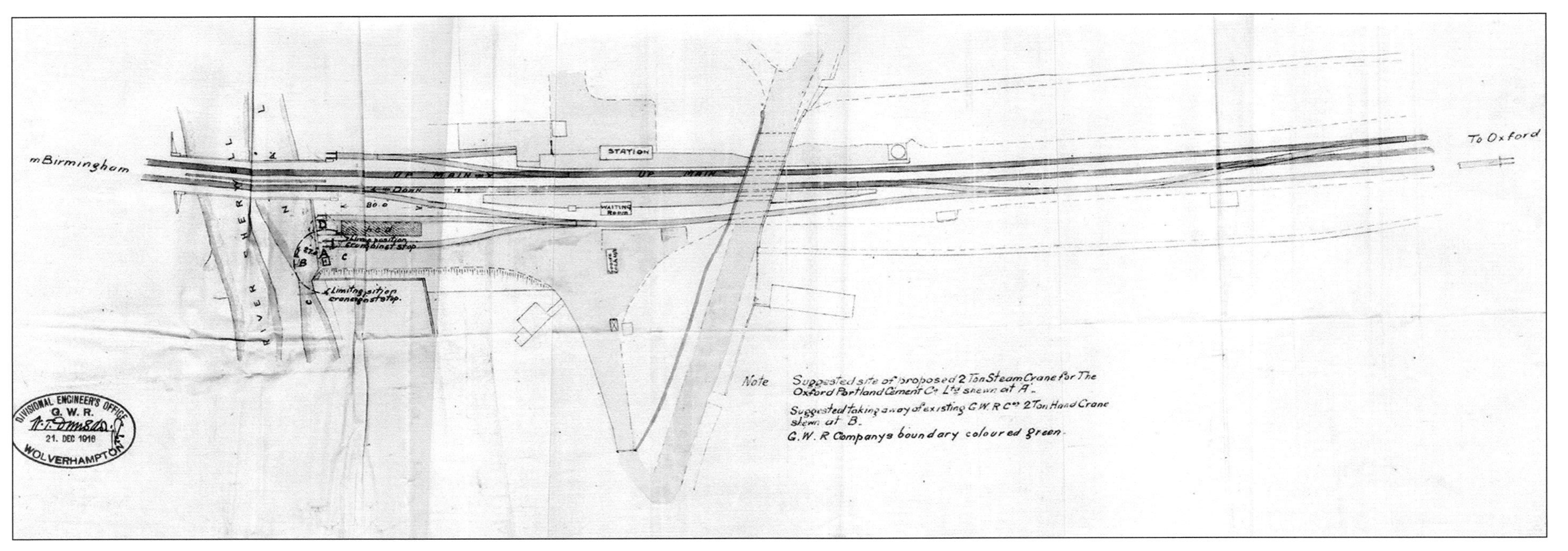

Figure 8: This very long original hand-coloured track plan from the GWR Wolverhampton drawing office shows the new sidings and the steam crane for unloading from the canal boats and to raise the cement bags on to the platform, where the cement store building was situated. It states 'taking out the 2-ton hand crane and replace with the steam crane' dated 1916.

Courtesy Chris Down collection.

The Great Western Railway Company and the Oxford Portland Cement Company Ltd. whose registered office is at Kirtlington; agreement as to a Steam Crane on the Great Western Railway at Bletchington Station in the County of Oxford. Subject to the Licensees obtaining any license that may be required and upon completion by the Licensees to the satisfaction of the Engineer of the Company; also of the strengthening of the canal retaining wall of the company's siding alongside the Oxford Canal, the company will permit the licensees at their own expense to erect a 2-ton steam crane at point 'A' and only use such crane for rail-borne traffic; rental for the easement £1 per annum; on completion of the crane, remove the existing hand crane; provided all rail-borne traffic arising at or destined for the licensees Works at Kirtlington near Bletchingdon shall be sent over the Company's Railway.

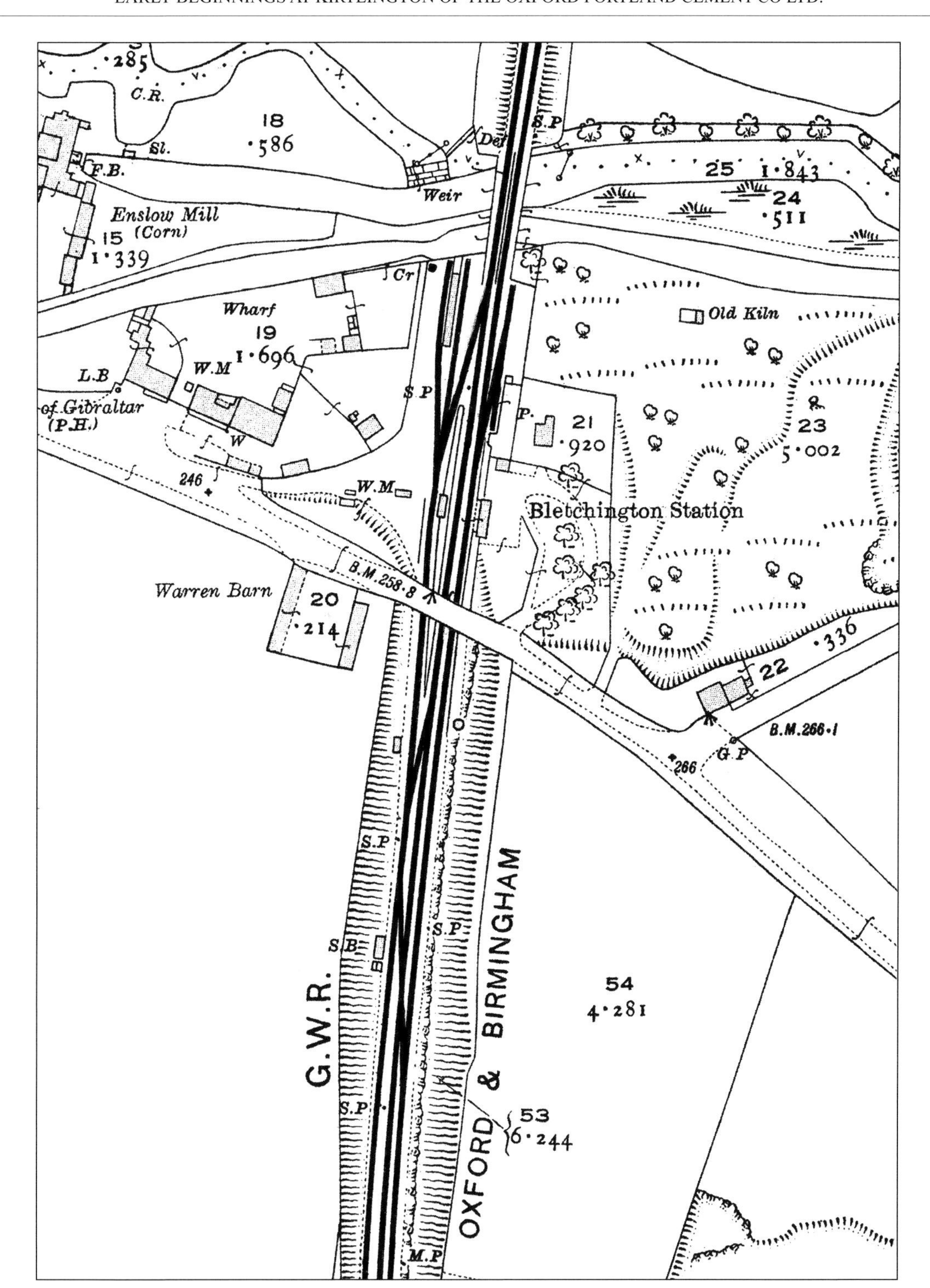

Figure 9: The two short sidings to hold 14 wagons with a platform inbetween at the northern end of the Bletchington GWR Station, (top) with a narrow lean-to warehouse on the platform was authorised on 25th of February 1909 (MT29/85/41) by the GWR.
The dot alongside these two sidings shows the position of the 2-ton crane (later a 2-ton steam crane) used to lift the cement bags from the Oxford Canal OPCC barges onto the wharf before they were transferred by hand into the store shed situated between the two sidings. No actual date as to when the GWR laid these two sidings is recorded, but they were in use by 1910. Note that this map shows the signal box south of the Enslow/Bletchington road bridge which can be seen in Plate 48.
In December 1914, at a cost of £854 to the GWR, a connection was made to the down refuge siding plus the repositioning of the station signal box as seen in the above map, to avoid the main line being used for shunting – this was all due to the vast increase in cement traffic.
Reproduced courtesy National Library of Scotland under the Creative Commons Attribution Licence.

Plate 17: Bletchington GWR Station (looking north to Birmingham). On the left are the two short sidings with the cement store in between (white shed, with a sloping roof) and just beyond that can be seen the 2-ton steam crane, used to winch the cement bags from the canal barges on to the wharf. It has been said that this store shed was constructed like this so that the weight would not damage the high canal retaining wall but the style is more likely to have been used as the width available was insufficient to construct a conventional brick building.
Oxfordshire County Council, Oxford History Centre; ref 1833402.

Plate 18: A poor-quality photograph showing these two short sidings for the Oxford Portland Cement Co. Ltd. transfer of cement from canal to rail. The transfer shed built in 1906, situated on the platform between the two sidings, was clad in an unusual covering of pressed steel sheets, shaped to look like lapped weather-boarding and produced by S. Taylor & Co, Lionel Street, Birmingham, from a patent submitted in 1905, which was granted on 2nd of August 1906.
Courtesy from the late J.H. Russell collection.

Plate 19: A further look in 1962 at the Bletchington lean-to style storage shed with the two short sidings one on either side. Note that the canal-side crane has disappeared by this time.

Courtesy Paul Karau collection, photographer P. Garland.

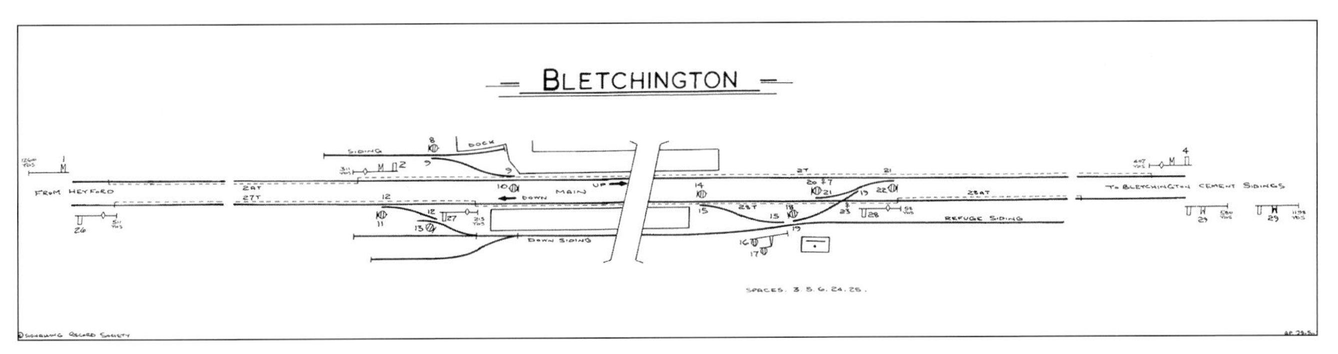

Figure 10: The Signalling Record Society diagram for Bletchington Station signal box including the two special cement usage sidings at the northern end of the station on the west side of the main line tracks.

Courtesy Chris Turner, Signalling Record Society.

Interestingly the Kirtlington cement works had no access by road or railway, even the small, very rough Mill Lane that ran alongside the works was unusable for any form of transportation other than limited use by horses. This was slightly improved around 1922, when a temporary steel ramp was made running into the works from this lane, so that the two Foden steam wagons that the company had purchased, could be used to deliver cement to local areas that could not be reached by rail or canal.

The works employees, right from the opening of the Kirtlington works, had to walk from Kirtlington village along a long footpath and then use a steep flight of rickety wooden steps leading down to the canal to get to the cement works *(see Plate 26)*.

Employees arriving from the west, had to cross the canal to get to work by using Pigeon Lock, which was some way from the cement works. However, if they could walk along the canal tow path to right opposite the works, then all they needed was a bridge across. The Oxford Canal Co. had been asked to help by constructing a bridge for the employees and even for the barge horses to cross the canal but the cost involved prohibited this. The canal company eventually agreed in 1911 however to operate two ferries across the canal at a fee of £1 per annum. One boat for the employees and the other for the horses! New stables were also needed for the many horses on operational duties and these were built next to the canal very near to the Kirtlington works but on the opposite bank alongside the towpath.

An interesting change in the Directors listed on the 1st of June 1915 were: – Dillon, Captain Arthur Henry at Barton Lodge, Steeple Aston, Oxford, stationed at Oxford; Holland, Captain, Harrington Inns of Court 27 Cannon Street London EC, stationed in France and Dillon, Mrs Hilda, Barton Lodge, Steeple Aston, Oxford; and that is all, so where have the others gone?

3

Names and addresses of the persons who are the Directors of the

THE OXFORD PORTLAND CEMENT Co Ltd
KIRTLINGTON, OXFORD. Limited,

on the first day of June 1915.

Names.	Addresses.
Dillon. Captain Arthur Henry. QOOH.	Barton Lodge. Steeple Aston. Oxford. Stationed at Oxford.
Holland. Captain Henry W. Inns of Court. [illegible]	27 Cannon Street. London. EC. Stationed in France.
Dillon. Mrs Hilda.	Barton Lodge. Steeple Aston. Oxford.

Figure 11: The page entry of the remaining OPCC directors by 1915.
Courtesy National Archives; ref – BT 31/17576/86120.

Much later, in 1920 and as stated before, a steep ramp had been constructed from Mill Lane when a huge Lancashire boiler was purchased which was too heavy and tall for canal transportation.

The Oxford Canal was the life blood for the works and all the initial construction building materials had to come in by the canal. The coal, sand and gypsum that were used at that time for cement making also arrived by this means.

The use of the Oxford Canal could not always be taken for granted and in the times of severe icing (and the canal ice-breaker from Heyford wharf was being used elsewhere), the boatmen were expected to stand at the bow of the canal boat and use a handmade long ash pole equipped with an iron tip and hook, that the OPCC blacksmith had made, to break up and push the large chunks of ice out of the way – a very hard task and not always successful so causing disruptive delays and even total stoppages.

The canal water depth also caused problems and the OPCC complained in December 1921 about the lack of dredging by the Oxford Canal Company – the cement company recorded in their minutes that 'little was done'. However, in the spring of 1923 an old steam dredger from 1911 was put to work between the Kirtlington works and Enslow Wharf. It raised over 70 tons of large stones, some weighing over 5cwt and over 22,000 tons of mud.

Plate 20: The completed Kirtlington cement works photographed sometime around the 1920s, showing the tramway tracks radiating out to the quarry face, situated off to the extreme right – note how the Oxford Canal banks in the distance are now heavily wooded, compared to Plate 13. It is possible to see some of the tramway skips standing at the entrance to the open shed just to its right of the big chimney. *Courtesy late Ted Thornton collection.*

The cement company, over the years, had purchased, in total, eight canal barges used mainly to move the cement from the works to Bletchington GWR Station rail head, and delivery to their yards in Oxford, Birmingham and Coventry. These barges all had names of local villages, except one with a city name! In 1908 '*Washford*' was the first boat purchased, followed in 1909 by canal barge '*Kirtlington*'; 1910 saw '*Bletchington*' followed by '*Tackley*' in 1911 *(Plate 22)*. Then a large gap, mainly due to the First World War, before in 1920, barge '*Rousham*' came online, followed by two more barges – '*Oxford*' and '*Heyford*'. The last canal boat purchased was '*Banbury*' in 1921.

Mr Tooley, of Banbury boat-builders, recalled that some of the last barges his company produced were for the Oxford Portland Cement Co. Ltd. He had delivered two much earlier but the third, which was in construction in 1928, was cancelled.

The OPCC barges painted livery was light blue for the framing, white panels with the barge's name in blue lettering and the canal official numbering etc. painted in black *(Plate 23)*. The barge's main hold was all covered with tarpaulin sheeting to keep the contents dry, as any wet coal delivered warranted an extra charge that was deducted by the OPCC from the barge captain's wages!

An article by Hugh Compton on the OPCC dated 1994, held at the Oxfordshire County Council, Oxford History Record office states:

> The canal boatmen who undertook this work for the OPCC, were well known 'Oxford' number ones, William Grantham of Banbury, with his boat *Hannah and The Cat* and his son George with his boat *Marlborough and Rose of Sharon*. Arthur Coles of Thrupp, accompanied by his mother, Louisa with their boats *Providence* and *Envy* (both registered at Oxford on the 29th January 1906 with boat numbers 79 and 80) also worked this trade before 1914. But after that they went to work for the Coop and then their boats were worked by Abel Skinner, the father of the well-known Joe Skinner.

Eventually, he bought both barges and had them re-registered at Banbury on 27th July 1920. Among other rare barge visitors to the cement works was barge *Gender*, which came in from Loughborough loaded with Gypsum.

As can be seen from the OS map in *Figure 2*, the materials from the quarry face were moved on a crude tramway system which was recorded to be about 2-foot gauge equipped with 'Jubilee Skips', capable of holding about a ton of material. These were pushed by hand to the works, recorded to be on a slightly down gradient for ease of movement! However, there is reference that occasionally horse power was used hired from a local farmer in lieu of payment for stone he needed for the farm. It is important to remind readers that at this time, this quarry work was very hard and a very dirty job!

For interest, a precis is now included to show the way the works operated in the manufacture of the portland cement, by the so called 'dry process.' The levels (called benches) that the quarry men had to work were of Cornbrash (clay), Lower Forest marble (hard limestone) and Great Oolite (cement stone) from the quarry face. The trucks of clay and limestone were pushed to the works and emptied by hand onto the clay rollers which then carried it to the blend mixers and onto the crushers. From there the materials passed into the 'raw' mill – a metal tube some 20 feet long and 6 feet in diameter. This tube was divided into three sections by internal plates that were perforated with 3/8ths inch holes, which allowed the ground down material to pass through. In the three compartments were a grinding agent, initially flint pebbles, later replaced by steel balls in the first two sections and, in the last section, by 'holpebs' (pebbles, later replaced by hollow metal balls). This tube was heated from the factory steam boilers which powered and heated parts of the works and consisted of two large Lancashire boilers which supplied steam to a Marshall horizontal condensing steam engine. The mechanical power was transmitted around the plant by 3-inch cotton ropes and belts which also drove a new-fangled electric light generator. Note that the water for the boilers came from the Oxford Canal and this cost the company £5 in the first year, but by 1910 the Oxford Canal Company had reduced this to just £1 per year.

The resultant ground-down material passed into a concrete hopper, whence it passed via a troughed screw feeder into a mixer fed with water to form nodules, this mix then fell back into a rotary kiln. This kiln being 60 ft long and again 6 ft in diameter, rotated slowly at 1 rpm and was driven by a ring gear around the outside of the kiln. Pulverised coal was then blown in from the bottom end of the kiln and fired, which produced a very high temperature causing the material to convert into a 'vitreous blue-black clinker.' Below this kiln, a 24 ft cooling cylinder was placed and from this cylinder, the clinker was weighed and gypsum was added to control the speed of the setting time. This material was then ground into small balls of cement and during this process, an enormous loss of heat went straight up the chimney causing clouds of dust to annoyingly be emitted, covering large areas around the Kirtlington village and surrounding district.

In the early days, this ground cement was stored in concrete bunkers and then hand filled into 2 cwt jute sacks which when emptied were returned to the factory for reuse. By 1915 a mechanical bagging machinery was installed (driven by compressed air) which could handle ten tons of cement per hour.

The Kirtlington quarry consisted of layers made up of 11 to 19 ft of clay, 30 to 33ft of hard limestone and 2½ ft to 4ft of cement stone. The quarrymen cleared off the top soil and then blasted the quarry face and the clay and limestone were then loaded separately into Jubilee Skips and pushed to the works. They were then fed in a ratio of five trucks of stone to two of clay – or as instructed by the works chemists, which could vary depending on the daily outside temperature.

Large amounts of coal were needed and came from all areas by canal, as rail transshipments in the early days were too costly. In 1914-15, an experiment was conducted, to determine which coal delivered had the best calorific value, with seven West Midlands coals taking part – the Newdigate, Charity and Exhall coals were tested for use as steaming coal for the kilns and Griff Rough Engine coals for the works boilers. The

Plate 21: A fine view, photographed in c1920 of the Kirtlington/Washford Pits works canal wharf that had been vastly improved around 1915. The unidentified canal barge is unloading coal, with the barge's family looking on and another boat behind awaiting its turn. *Authors collection; courtesy late J.H. Russell collection.*

Plate 22: One of Oxford Portland Cement Company's own canal barges named TACKLEY, purchased in1911, is seen with a happy female crew member and some female cement workers on board. It is probably at the canal winding hole at the Kirtlington/Washford cement works or possibly at Enslow wharf at Bletchington GWR Station on the Oxford Canal. The photograph date is said to be 1911 and one wonders if the 'bagging' ladies from the cement works were being taken for a dinnertime, summer's joyride on the newly purchased barge?
Authors collection; courtesy late J.H. Russell collection.

Plate 23: This fine colour illustration is of TACKLEY, one of the OPCC canal barges.
Kind permission of Waterways World magazine.

results were that Warwickshire Second Steam coal from the Pooley Mine at 12s 5d per ton, then Cannock Chase at 17s 3d per ton were deemed the best.

During the First World War, with male personnel shortages, three local women from Tackley village (including Mrs Dillon, wife of the now Major Dillon) were employed to clear the top soil of the quarry edge using spades and strong (works made) elm wheelbarrows – a extremely dangerous job being so near to the quarry edge! Women were also used in the bagging plant but the works managed to keep around seventy male employees that were exempt from wartime army service duties. The works output in November 1915 was down to 967 tons of cement, with a recorded 268 tons going for 'secret Government Work'.

It is interesting to see how the canal was coping, as the coal was coming in by two boats paired-up giving 52 tons total (26 tons per boat) plus the input of gypsum and sand, not forgetting the enormous output of cement going out! So, the 'toing and froing' from the GWR Bletchington railhead and the output from the works wharf were also very busy. Cement to the Oxford sales yard in Worcester Street (telephone number 185) by canal barge took a day, however, cement delivery to the Coventry canal wharf yard (telephone 38), took five days and Birmingham, a long and tedious eight days.

The preferred coal for pulverising came in from the Pooley Mine in Warwickshire at a cost per boat of £6 and, in the 1920s, seven boatmen served the works averaging seven to eight round trips per month. Coal was unloaded by hand, using shovel and wheelbarrows and four men would unload the two barges carrying 52 tons, between 7am and 4pm daily with only a 30-minute lunch break allowed.

As previously stated, in 1925 two modern Foden steam lorries were purchased to help with local deliveries and these were passed on to the new cement works at Shipton-on-Cherwell, whose history and operation will be dealt with in later chapters.

The factory (known locally as just 'the works') production proved costly during the General Strike of 1926 and the coal needed became expensive and extremely hard to find, so the writing was on the wall for the Kirtlington (Washford Pits) cement works to close. The output thus lost amounted to over a third of production. By the end of 1928, with the plant worn out, the quarry virtually exhausted, the road access still very dismal, plus the new Shipton cement works nearly ready to go into production, the outlook was grim. The decision was made in early 1929, when under the banner of the new company Oxford & Shipton Cement Ltd. (but still controlled by A.H. Dillon) – the directors decided to close the Kirtlington cement complex for good.

Plate 24: A look at the cement works wharf area of the Kirtlington/Washford Pits works in the 1990s, after the area was returned to a natural habitat site alongside the Oxford Canal. Note the state of the towpath on the left hand side of the photograph compared with the modern image below.
Courtesy Oxfordshire County Council, Oxford History Centre.

Plate 25: A view of the Oxford Canal on 11th of March 2010 with the derelict Kirtlington cement works canal wharf on the right. Photograph by Mike Todd and reproduced under the Creative Common Licence.

Plate 26: The two views of the Kirtlington quarry as seen today, now designated as an important SSSI. The left image shows the quarry face. *Authors' collection purchased from an international auction site.*
The right image features the steps down into the quarry floor (which the workers from the local area would have descended to reach the factory situated next to the canal).
Photograph by Jim Thornton, reproduced under the Creative Common Licence.

When finally abandoned in 1932, nearly the whole Kirtlington cement works was sold for scrap to the local merchants, Friswell and Sons, Banbury, Oxfordshire but the large steam engine went to an Indian cotton mill and worked there until 1936. The cement mills at the works were made by the German company Friedrich Krupp of Essen and one of the gear wheels from the cement mill was purchased in 1929 for use in the operating mechanism of the 16 ft water wheel of 'Flight Mills'; later being built into the fabric of an old house bearing that name in Kirtlington village.

Lastly, it is said that first two bags of cement from the cement works were sold and used for coping work on the wall around the Kirtlington old vicarage.

Back to 'Flights Mill,' the then owner, Laurie Giles reportedly still had several receipts for cement bought by his family from the Kirtlington works, one of which was signed 'F. Adams – Cashier,' who was thought to have been a relative of one of the cements works employees, but later was found to be the daughter of the police sergeant at Kirtlington village.

After the 'official' closure of the cement works in 1929, a smaller mill and brand-new crusher plant was purchased by a new company Oxford & Shipton Cement Ltd. and was set up on the quarry floor. It was used to quarry small amounts of limestone, which was ground up for chicken grit and small size roadstone, but this work did not last long and the end came in 1932 when the site was abandoned and then silence (and above all clean air) finally descended on the area and nearby Kirtlington village once more.

See Appendix One: For details of cement production totals, wages paid to the employees and days lost in production at Kirtlington Cement works.

Authors' note:

Several of the events in the following Chapter Three happened roughly at the same time and therefore there could have been several ways to have recorded them. We hope that our interpretation and the historical sequence makes the best sense of these. It might seem a little disjointed, as the implication was that the impending closure of the Kirtlington cement works was a surprise; whereas it seems that Arthur Dillon, at least had personally seen it coming some years before hence the purchase of land at Shipton-on-Cherwell under the guise of his new short-lived company, Shipton Syndicate – this no doubt to make himself a nice personal profit when he sold the land later to OPCC for its continued existence.

Information regarding the geology at the Shipton quarry

Cement works are usually established in locations where the local geology can provide the raw materials needed. The main ingredients required are chalk or limestone (to supply the calcium) and clay or shale (the alumina and silica), the complete raw mix needing these in the proportions 3 or 4 to 1.

In most cases, cement works have two separate quarries, one for the chalk/limestone and the other for clay/shale. However, Kirtlington and Shipton were among the minority of locations where all the raw materials could be obtained from the same quarry because, in effect, nature had already mixed them. However, they did not occur in the precise proportions required and so, within the single quarry, several different levels and materials had to be worked simultaneously to blend the various areas into the correct proportions. To achieve that, the quarry railways had to serve several different working areas and, at Shipton, usually at least two trains and two excavators were needed.

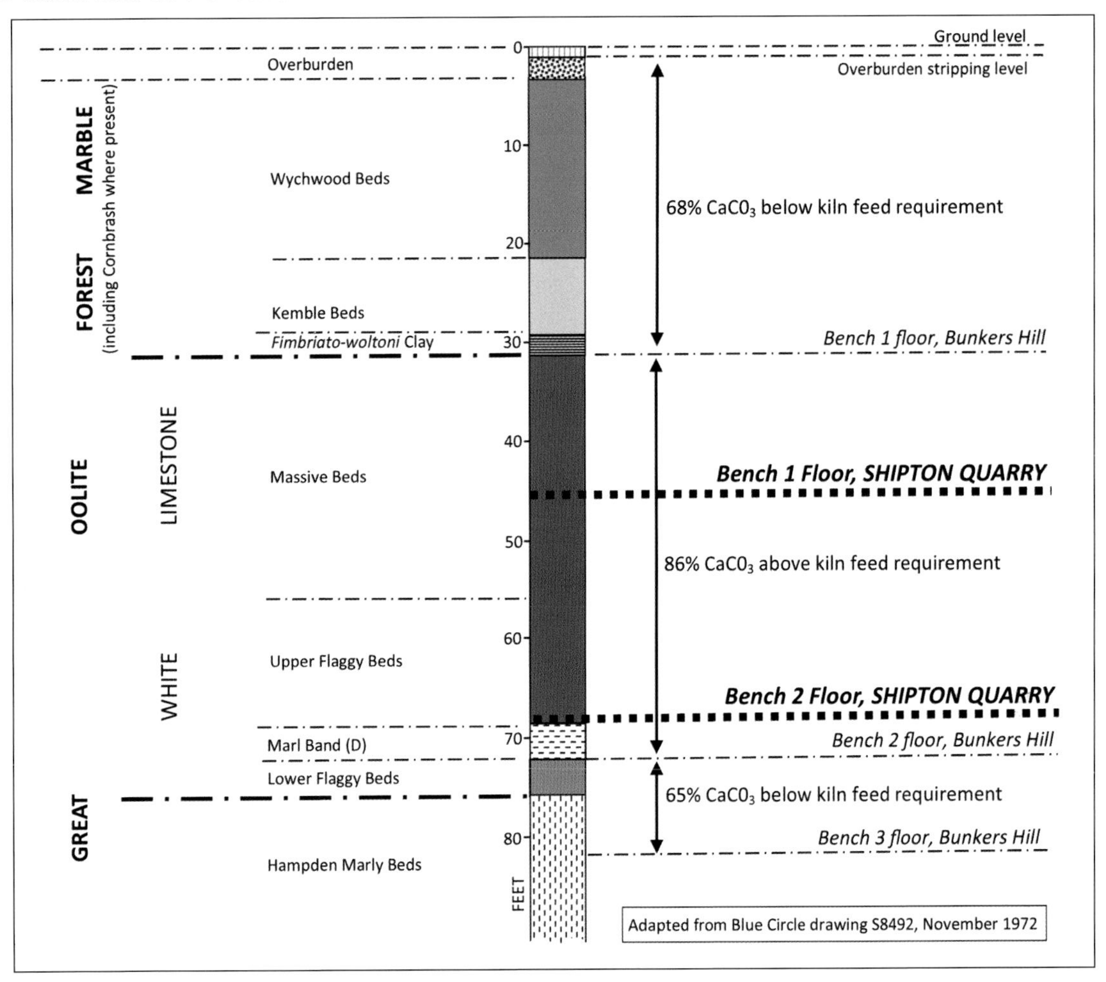

Figure 13: The manufacture of cement requires a calcium carbonate ($CaCO_3$) content between 76 and 79%. Few if any natural rocks meet that specification, so that blending of different materials is required. With pure chalk or limestone and clay or shale, blending is usually simple. At locations such as Shipton, the strata are more complex and so too is the blending.
This geological section is diagrammatic but shows how the original Shipton quarry was dug in two benches (levels), the upper bench generally containing too little calcium carbonate and the lower bench too much, the two being blended by quarrying them separately and then controlling how they were mixed, with a final 'fine tuning' after the crushing and blending stage. The drawing was in fact prepared by APCM to decide how best to work the Bunkers Hill land (which never happened) but it was envisaged as three benches, the upper two not being dissimilar to Shipton's existing quarrying, but with a third bottom bench of poorer quality material which could just be blended in. Shipton works had trialled this bottom bench over a small area in the northeast corner of the quarry to prove its practicality. Note that the geological descriptions are those of the 1970s era; in recent years the terminology has changed. *Diagram drawn by Chris Down.*

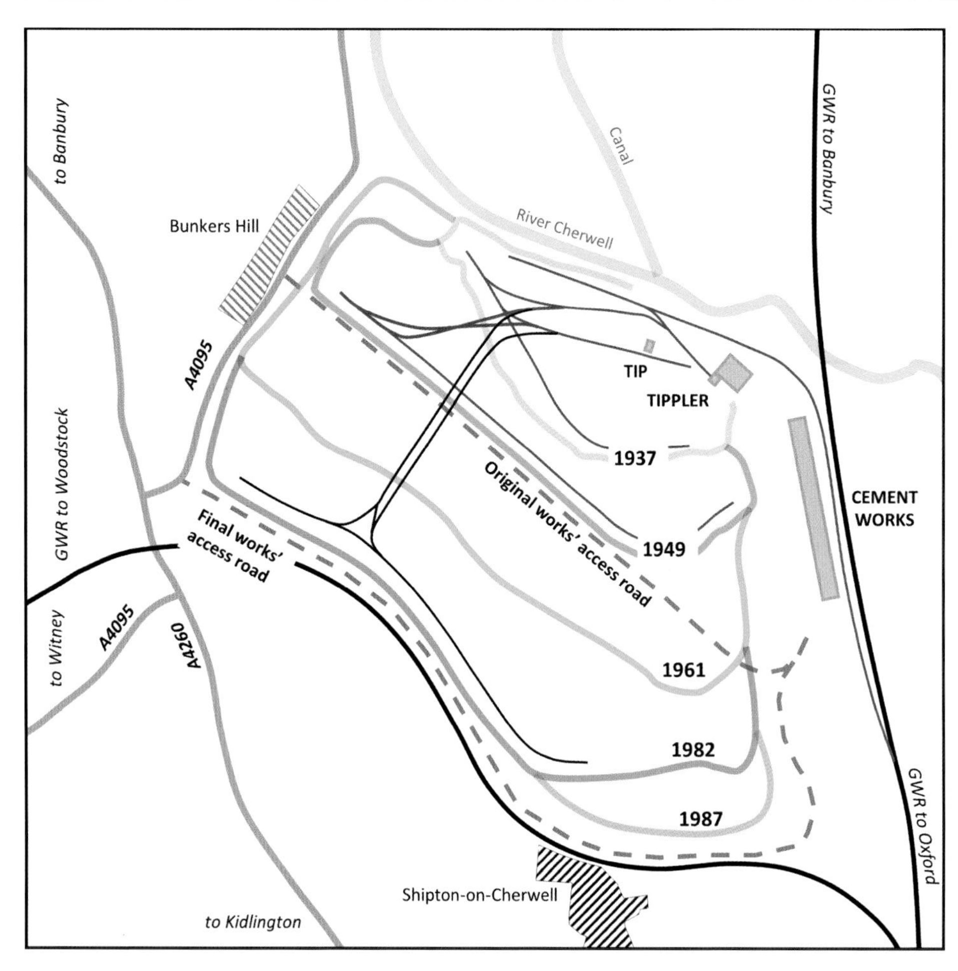

Figure 14: This diagrammatic sketch map illustrates the overall development of the Shipton quarry from its inception to closure. The various coloured lines indicate the rough extent of quarrying at the dates stated. The works internal private railways are shown in thin red (as they were laid pre-1949) and the thin black lines (as the tracks were extended post-1949). These are purely diagrammatic, showing just general routing and no details of track layouts, which changed quite frequently and are mostly unknown to any accuracy.

Diagram drawn by Chris Down.

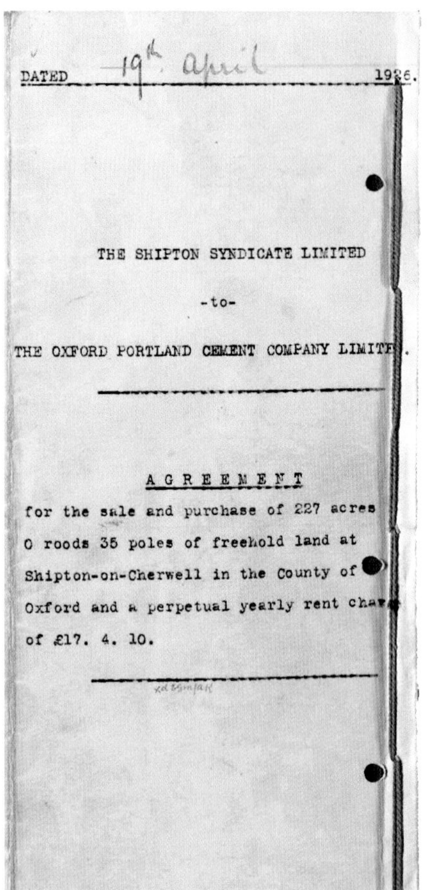

DATED 19th April 1926.

THE SHIPTON SYNDICATE LIMITED

-to-

THE OXFORD PORTLAND CEMENT COMPANY LIMITED.

AGREEMENT

for the sale and purchase of 227 acres 0 roods 35 poles of freehold land at Shipton-on-Cherwell in the County of Oxford and a perpetual yearly rent charge of £17. 4. 10.

Anticipating the closure of Washford Pits (Kirtlington) cement works due to the quarry becoming exhausted, chairman A.H. Dillon gathered a formidable selection of gentlemen to be the directors of a new company. Its purpose would be to take over the OPCC as a going concern and to develop the lands purchased via an agreement of the 19th of April 1926 between OPCC and the Shipton Syndicate, on which a new cement works and associated quarries were proposed and which was expected to produce 125,000 tons of cement a year.

The agreement of the 19th of April is reproduced in *Plate 28* and following pages. The sale price for 227 acres 35 poles of land was £17,000 0s 0d, with a deposit of £300 0s 0d. The deal was completed on the 8th of May 1926. Then it states that the final balance of the purchase price shall be satisfied as by two thousand seven hundred pounds in cash and fourteen thousand pounds of Ordinary £1 shares in the Capital of the Oxford Portland Cement Co. Ltd. with shares numbered 29,206 to 43,205 inclusive.

Plate 28: Left and the following pages are the 1926 agreement between the Shipton Syndicate Limited and the Oxford Portland Cement Company Limited.

Courtesy National Archives; ref – BT 31/17576/86120.

86120

REVENUE ROYAL COURTS OF JUSTICE TEN SHILLINGS

COMPANIES REGISTRATION FIVE SHILLINGS

AN AGREEMENT made the nineteenth day of April One thousand nine hundred and twenty-six BETWEEN THE SHIPTON SYNDICATE LIMITED whose Registered Office is at 3 and 4 Clements Inn Strand in the City of Westminster (hereinafter called the Vendors) of the one part and THE OXFORD PORTLAND CEMENT COMPANY LIMITED whose Registered Office is at Kirtlington in the County of Oxford (hereinafter called the Purchasers) of the other part WHEREBY IT IS AGREED as follows:-

1. The Vendors will sell and the Purchasers will purchase at the price of Seventeen thousand pounds FIRSTLY ALL THOSE pieces of land and land covered by water situate in the Parish of Shipton-on-Cherwell in the County of Oxford containing in the whole 227 acres 35 poles or thereabouts more particularly delineated and described in the Plan drawn upon an Indenture of Conveyance dated the twenty-third of March One thousand nine hundred and twenty-one and made between The Wardens and Scholars of Saint Mary College of Winchester in Oxford commonly called New College in Oxford of the one part and the Vendors of the other part and therein edged with the colour pink TOGETHER with such right and liberty as hitherto enjoyed by the Vendors and their predecessors in title (in common with all other persons (if any) who have or may hereafter have the like right) at all times and for all purposes either with or without horses carts waggons and any other vehicles to use the road coloured brown between the points marked A and B on the said plan AND SECONDLY ALL THAT perpetual yearly Rent Charge

REGISTERED 86538 28 JUN 1926

Filed by Cox & Hardale 7 Tower Royal EC4

280

of Seventeen pounds four shillings and ten pence charged upon and issuing out of the plot of land and land covered by water containing together five acres one rood or thereabouts(being the towing path and half the bed of the River) delineated and coloured blue on the said Plan and all future payments of such Rent Charge TOGETHER with all powers and remedies for recovering and compelling payment of the same AND the full benefit of all provisions relating thereto contained or referred to in the Acts of Parliament 9 Geo.III Ch.70 and 10 Geo.IV Ch.48 and the Acts amending the same under or by virtue whereof the said land and land covered by water coloured blue on the said Plan are now vested in the Oxford Canal Company subject to the payment of the said Yearly Rent Charge.

2. The Purchasers have paid to the Vendors' Solicitor the sum of Three hundred pounds by way of deposit and in part payment of the purchase price and the purchase shall be completed on the eighth day of May One thousand nine hundred and twenty-six at the office of the Vendors' Solicitor Mr. F.R.Allen 3 and 4 Clements Inn Strand London W.C.2.

3. The said balance of the purchase price shall be satisfied as to Two thousand seven hundred pounds by payment in cash and as to Fourteen thousand pounds by the allotment to the Vendors or their nominees of Fourteen thousand Ordinary Shares of One Pound each in the Capital of The Oxford Portland Cement Company Limited credited as fully paid up and numbered 29206 to 43205 inclusive.

7

Interest (if any) shall be payable only on the cash balance of the purchase price.

4. The Vendors will sell and the Purchasers will purchase the farming implements stocks crops and produce upon the said premises at a valuation as on the sixth day of April One thousand nine hundred and twenty-six to be made by Messrs Franklin & Jones of Oxford on behalf of both parties.

5. On payment of the balance of the purchase money and of the amount of the valuation and allotment of the said shares in manner hereinbefore provided the Purchasers shall be entitled to vacant possession of the property.

6. The title to the property shall commence with the said Indenture of Conveyance dated the twenty-third of March One thousand nine hundred and twenty-one coupled with a Statutory Declaration by the Bursar of New College aforesaid dated fourteenth March One thousand nine hundred and twenty-one and the Purchasers shall not be entitled to make any requisition or objection in respect of the prior title to the property which formed part of the ancient estates of New College Oxford.

7. The property is sold subject to the General Conditions of One thousand nine hundred and twenty-five so far as the same are applicable to a sale by private treaty and are not varied by this Agreement.

IN WITNESS whereof the Vendors and the Purchasers have caused their Common Seal to be hereunto affixed the day and year first before written.

THE COMMON SEAL of The Shipton Syndicate Limited was hereunto affixed in the presence of

A. Dillon
H.T. Wilson
Directors.

F R Allen
Secretary.

THE COMMON SEAL of The Oxford Portland Cement Company Limited was hereunto affixed in the presence of

A. Dillon
Hilda Dillon.
Directors.

Douglas [illegible]
Secretary.

Received the within mentioned deposit of Three hundred pounds this 19th April 1926

F R Allen

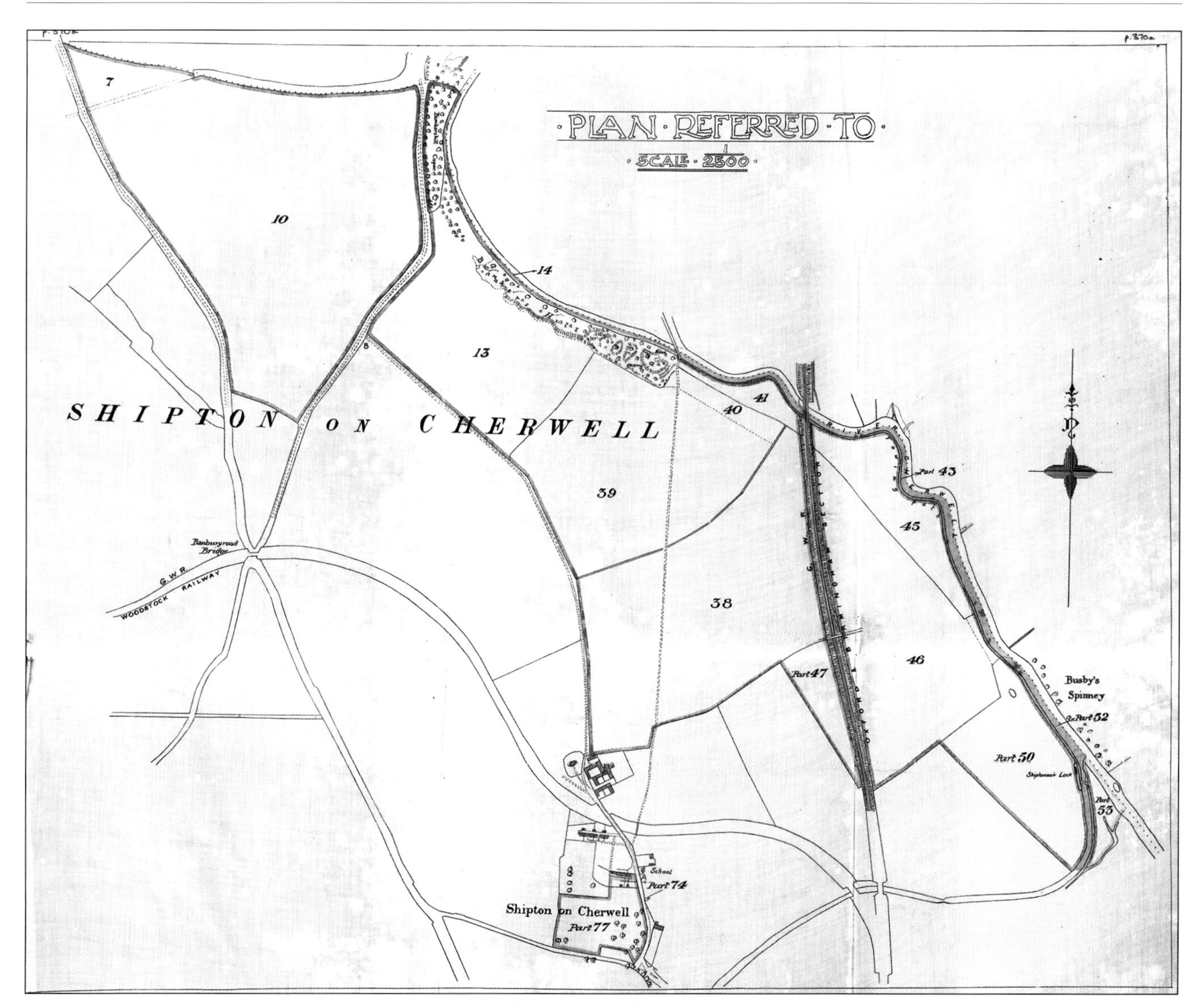

Figure 15: The accompanying plan is from a large folded map in the New College, Oxford University library archives and shows clearly, on the left of the map, the dangerous four-way road junction right under the GWR Woodstock branch line bridge – one road on one side of the bridge and the other, the other side. This was later amended to a staggered junction to avoid accidents due to the heavy traffic from the Shipton cement works. The mauve lines indicate the areas of land sold. *Courtesy New College Library Archives.*

The interesting agreement is included in full, with the hand drawn map above that shows the land acquired. This syndicate seems to have been in existence since the conveyance of the land between them and New College in 1921.

At this time the Oxford Portland Cement directors were now stated as Arthur Dillon, Hilda Dillon (who joined in 1915), Captain Holland, plus an unnamed Frenchman. It is presumed that sometime later, this Shipton Syndicate Limited would then sell this land as a separate deal when the Oxford Portland Cement Co. Ltd. was sold, earning Arthur Dillon and family a large return, however this was not the case, as in just a few days, on the 27th of May 1926, the Syndicate sent a letter to OPCC requesting that the share allocation in the New College agreement be dealt with immediately *(see Plate 29).*

Plate 29: The following two official documents reveal the distribution of the shares that the Shipton Syndicate Ltd. had required, with the last entry showing how many Arthur Dillon had allocated to himself (the largest) 4667.
Courtesy National Archives; ref – BT 31/17576/8612.

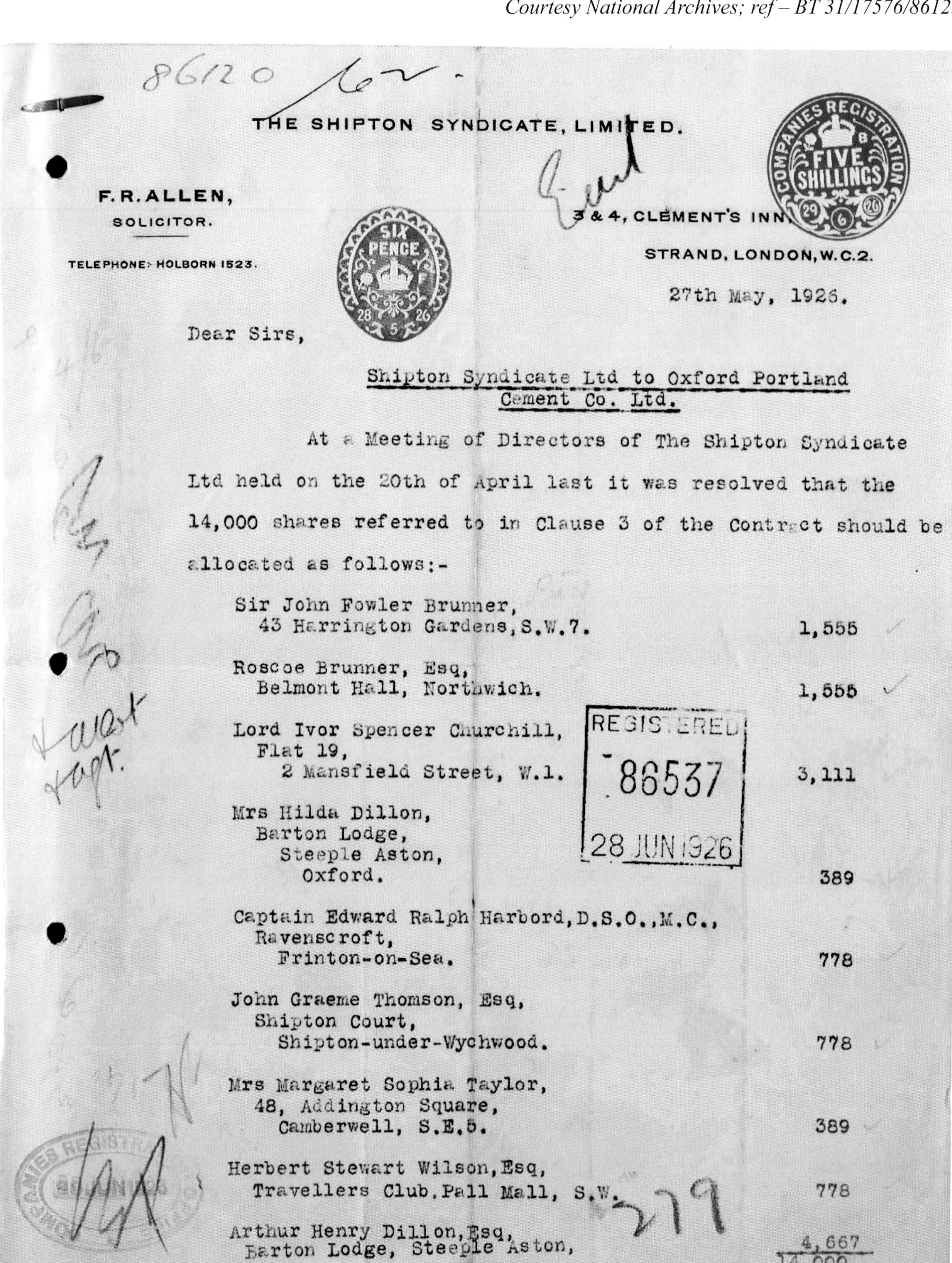

86120

THE SHIPTON SYNDICATE, LIMITED.

F. R. ALLEN,
SOLICITOR.

TELEPHONE: HOLBORN 1523.

3 & 4, CLEMENT'S INN,
STRAND, LONDON, W.C.2.

COMPANIES REGISTRATION FIVE SHILLINGS

SIX PENCE

27th May, 1926.

Dear Sirs,

Shipton Syndicate Ltd to Oxford Portland Cement Co. Ltd.

At a Meeting of Directors of The Shipton Syndicate Ltd held on the 20th of April last it was resolved that the 14,000 shares referred to in Clause 3 of the Contract should be allocated as follows:-

Sir John Fowler Brunner, 43 Harrington Gardens, S.W.7.	1,555
Roscoe Brunner, Esq, Belmont Hall, Northwich.	1,555
Lord Ivor Spencer Churchill, Flat 19, 2 Mansfield Street, W.1.	3,111
Mrs Hilda Dillon, Barton Lodge, Steeple Aston, Oxford.	389
Captain Edward Ralph Harbord, D.S.O., M.C., Ravenscroft, Frinton-on-Sea.	778
John Graeme Thomson, Esq, Shipton Court, Shipton-under-Wychwood.	778
Mrs Margaret Sophia Taylor, 48, Addington Square, Camberwell, S.E.5.	389
Herbert Stewart Wilson, Esq, Travellers Club, Pall Mall, S.W.	778
Arthur Henry Dillon, Esq, Barton Lodge, Steeple Aston, Oxford.	4,667
	14,000

REGISTERED
86537
28 JUN 1926

279

Number of Certificate } 86120.

Form No. 45.

COMPANIES ACTS 1908 TO 1917.

RETURN OF ALLOTMENTS from the 4th of June 1926

to the 18th of March 1927 of

See note (1) below.

OXFORD PORTLAND CEMENT COMPANY Limited.

Pursuant to s. 88 (1) *of the Companies* (*Consolidation*) *Act* 1908.

Companies Registration Office — FIVE SHILLINGS — Companies Registration fee stamp must be impressed here.

* Distinguish between Preference, Ordinary, &c.

*Number of the Shares allotted payable in cash ...... NONE

" " " " ......

Nominal amount of the Shares so allotted..............

" " " ....

Amount paid or due and payable on each such Share

" " " "

Number of Shares allotted for a consideration other than cash 5,000

Nominal amount of the Shares so allotted.............. 5000

Amount to be treated as paid on each such Share ... £1

† The contract constituting the title of the allottee must also be filed. See sec. 88 (1) (B) of the Companies (Consolidation) Act 1908.

†The consideration for which such Shares have been allotted is as follows:—

Services rendered by the Allottee to the Company.

38870

24 MAR 1927

NOTE.—This return must be filed within one month after the allotment is made, or in default a penalty of £50 per day is incurred. See section 88 (2) of the Companies (Consolidation) Act 1908.

(1) When a return includes several Allotments made on different dates, the dates of only the first and last of such Allotments should be entered, and the registration of the return should be effected within one month of the first date.

When a return relates to one Allotment only, made on one particular date, that date only should be inserted, and the spaces for the second date struck out and the word "made" substituted for the word "from" after the word "Allotments."

Presented for filing by

MONTAGU'S AND COX & CARDALE,
86 & 88, Queen Victoria Street, E.C.4.

Companies Registration Office 24 MAR 1927

The Solicitors' Law Stationery Society, Limited, 22 Chancery Lane, W.C.2, 27 & 28 Walbrook, E.C.4, 49 Bedford Row, W.C.1, 6 Victoria Street, S.W.1, 15 Hanover Street, W.1, and 66 St. Vincent Street, Glasgow.

PRINTERS AND PUBLISHERS OF COMPANIES' BOOKS AND FORMS.

Companies Form 6J.—6763-24.2.26 W134

166

[P.T.O.

Plate 30: These were most of the men that left the Kirtlington cement works when it closed to go over to the new company – the Oxford & Shipton Cement Ltd. and some were probably lucky to get a new house at Bunkers Hill. All were reported to have had over 27 years of service, and some even over 40 years.
Courtesy the 'Blue Circle' (the house magazine of the Blue Circle Cement) Volume 10, No. 1 (January 1956).

The final voluntary winding up of the original Oxford Portland Cement Company Limited (number 86120) was passed on the 3rd of March 1927, confirmed on the 18th of March 1927, and signed by Arthur Dillon on the 28th day of March 1927 – the liquidator being Arthur Taylor, a Bristol Chartered Accountant. The final winding up return was registered on the 14th of January 1928.

The new company offered positions at the new Shipton works to former Kirtlington employees and it is recorded that 47 of the Kirtlington workforce transferred over.

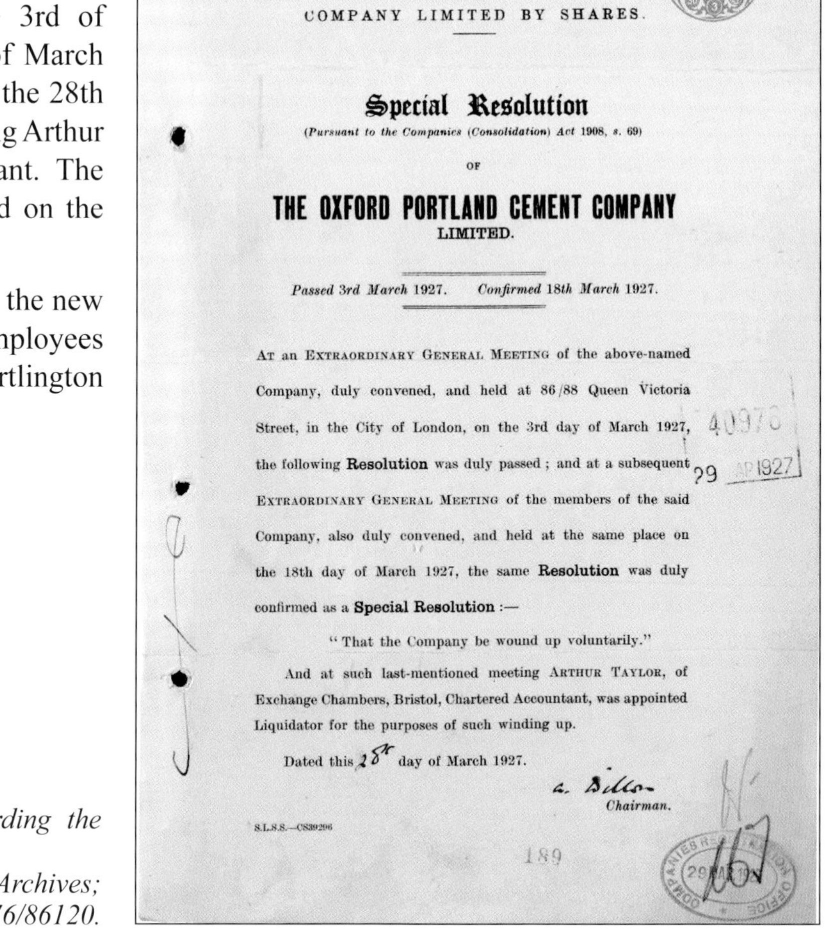

86120/67

The Companies Acts 1908 *to* 1917.

COMPANY LIMITED BY SHARES.

Special Resolution

(*Pursuant to the Companies (Consolidation) Act* 1908, *s.* 69)

OF

THE OXFORD PORTLAND CEMENT COMPANY LIMITED.

Passed 3rd March 1927. *Confirmed 18th March* 1927.

At an Extraordinary General Meeting of the above-named Company, duly convened, and held at 86/88 Queen Victoria Street, in the City of London, on the 3rd day of March 1927, the following **Resolution** was duly passed; and at a subsequent Extraordinary General Meeting of the members of the said Company, also duly convened, and held at the same place on the 18th day of March 1927, the same **Resolution** was duly confirmed as a **Special Resolution**:—

"That the Company be wound up voluntarily."

And at such last-mentioned meeting Arthur Taylor, of Exchange Chambers, Bristol, Chartered Accountant, was appointed Liquidator for the purposes of such winding up.

Dated this 28th day of March 1927.

A. Dillon
Chairman.

S.L.S.S.—CS39296

Plate 31: The Special Resolution regarding the winding up voluntarily of the OPCC.
Courtesy National Archives; ref – BT 31/17576/86120.

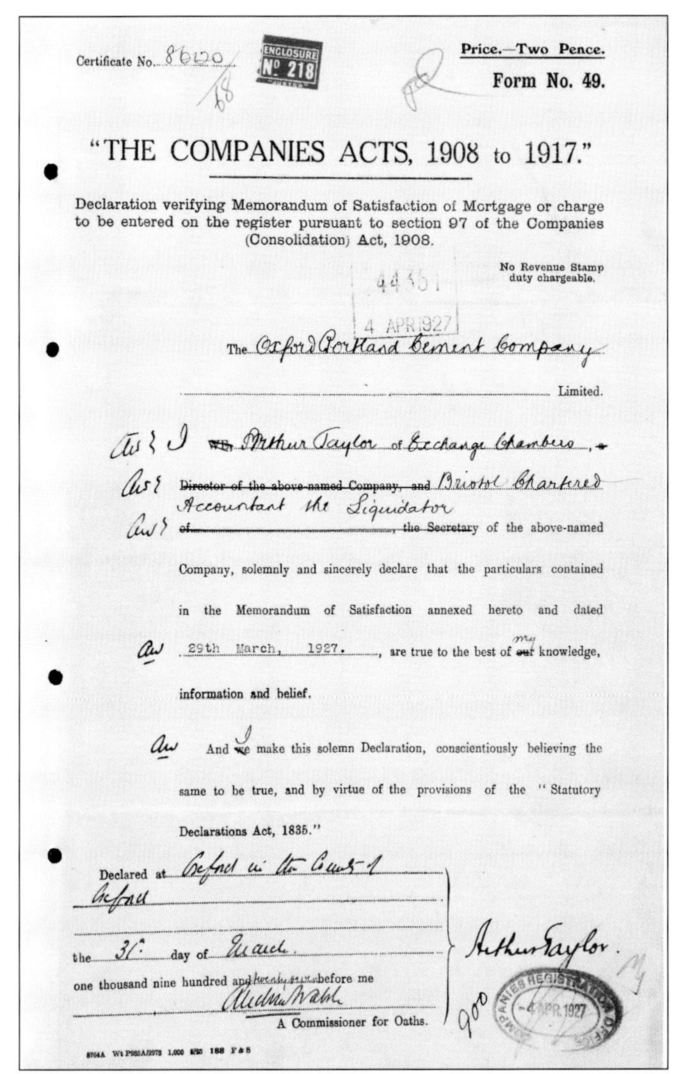

Certificate No. 86120 / 18

ENCLOSURE No 218

Price.—Two Pence.

Form No. 49.

"THE COMPANIES ACTS, 1908 to 1917."

Declaration verifying Memorandum of Satisfaction of Mortgage or charge to be entered on the register pursuant to section 97 of the Companies (Consolidation) Act, 1908.

No Revenue Stamp duty chargeable.

44351 4 APR 1927

The Oxford Portland Cement Company Limited.

I Arthur Taylor of Exchange Chambers, ~~Director of the above-named Company, and~~ Bristol Chartered Accountant the Liquidator ~~of ..., the Secretary~~ of the above-named Company, solemnly and sincerely declare that the particulars contained in the Memorandum of Satisfaction annexed hereto and dated 29th March, 1927., are true to the best of my knowledge, information and belief.

And I make this solemn Declaration, conscientiously believing the same to be true, and by virtue of the provisions of the "Statutory Declarations Act, 1835."

Declared at Oxford in the County of Oxford the 31st day of March one thousand nine hundred and twenty seven before me

A Commissioner for Oaths.

Arthur Taylor.

Companies Registration Office 4 APR 1927

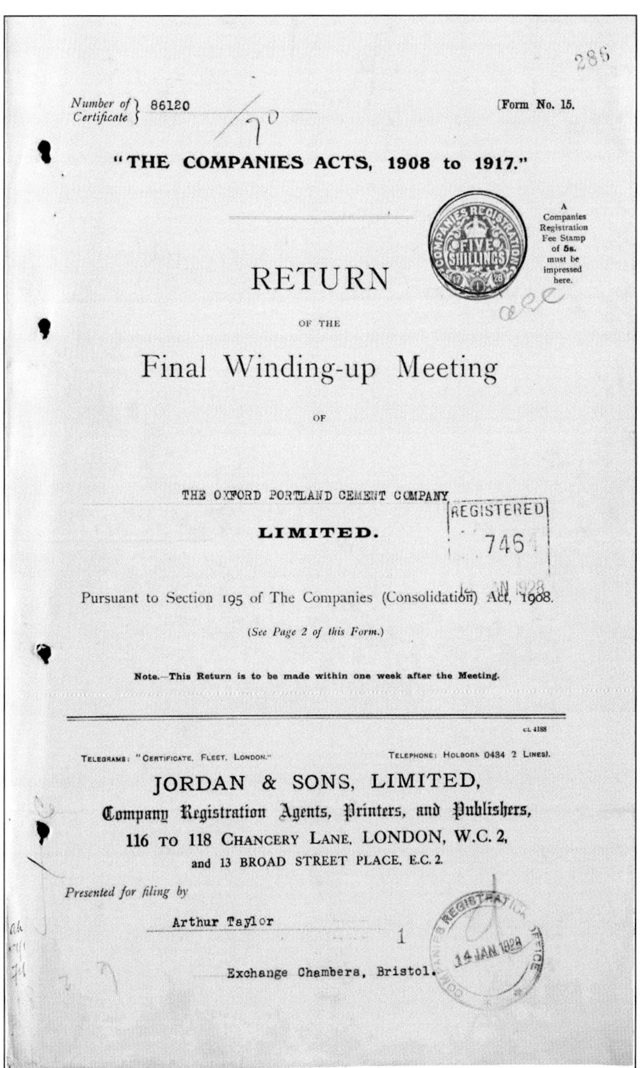

286

Number of Certificate 86120

Form No. 15.

"THE COMPANIES ACTS, 1908 to 1917."

A Companies Registration Fee Stamp of 5s. must be impressed here.

RETURN

OF THE

Final Winding-up Meeting

OF

THE OXFORD PORTLAND CEMENT COMPANY

LIMITED.

REGISTERED 7454

Pursuant to Section 195 of The Companies (Consolidation) Act, 1908.

(See Page 2 of this Form.)

Note.—This Return is to be made within one week after the Meeting.

Telegrams: "Certificate, Fleet, London." Telephone: Holborn 0434 (2 Lines).

JORDAN & SONS, LIMITED,

Company Registration Agents, Printers, and Publishers,

116 to 118 Chancery Lane, London, W.C. 2,

and 13 Broad Street Place, E.C. 2.

Presented for filing by

Arthur Taylor

Exchange Chambers, Bristol.

Companies Registration Office 14 JAN 1928

Plate 32 (left): The Memorandum of Satisfaction dated 29th of March 1927.

Courtesy National Archives; ref – BT 31/17576/86120.

Plate 33 (right): The final document of the Final Winding-up Meeting dated 14th of January 1928.

Courtesy National Archives; ref – BT 31/17576/86120.

The statutory Meeting for the new company (including the adoption of the new title) was held on the 3rd of May 1927 at Winchester House, Old Broad Street, London EC with A.H. Dillon as Chairman:

> Ladies and Gentleman,
>
> I am glad to have this first meeting of our shareholders. I must report that the purchase of the Oxford Portland Cement Co. Ltd. had now been completed and until April the 23rd, when the Kirtlington works closed for urgent repairs, they were working at full capacity. It will be the 8th of May that manufacturing will be resumed.
>
> The new Shipton quarry site on which the new works will be erected, has been surveyed by Maxted and Knott (the Company's consulting engineers) and the work of clearing the site was begun on February 21st. The layout of the new works has been so arranged that from the crushing of the raw materials to the dispatch of the finished product by road, rail or canal, there follows a continuous sequence of operations.
>
> Negotiations with the Great Western Railway for the construction of the railway sidings at Shipton are proceeding and it is expected that the actual construction will be commenced

within the next fortnight. Invitations to tender for the most important items of the plant required have also been sent out.

Provided that there are no industrial troubles, it is expected that the new works will be completely erected and produce cement in 15 months' time. I hope to be able to report good progress when the first annual general meeting of the company is held.

I now move that the report be adopted and will ask Sir John Brunner kindly to second the resolution.

Sir John Brunner seconded the resolution and it was carried unanimously.

The new company was to be called the Oxford & Shipton Cement Limited and was filed with the Register of Joint Stock and released to the public on Monday 7th of February 1927, with shares comprising, Government guaranteed loan capital of £100,000 and 400,000 ordinary £1 shares.

A Copy of this Prospectus has been filed with the Registrar of Joint Stock Companies.

The Subscription List will be opened on Monday, the 7th February, 1927, and closed on or before Wednesday, the 9th February, 1927.

Application will be made to the Committee of the London Stock Exchange for permission to deal in the Shares of the Company after allotment, and for an official quotation.

OXFORD & SHIPTON CEMENT Limited

(Incorporated under the Companies Acts, 1908 to 1917.)

SHARE CAPITAL - - £500,000.

DIVIDED INTO

500,000 Ordinary Shares of £1 each.

50,000 Shares are to be issued as fully paid as mentioned below, and 50,000 Shares are reserved for future issue.

LOAN CAPITAL £100,000 First Mortgage Debenture which His Majesty's Government is Guaranteeing under the Trades Facilities Acts.

ISSUE OF

400,000 ORDINARY SHARES OF £1 EACH AT PAR

PAYABLE:

2/6 per Share on Application.
2/6 per Share on Allotment.
5/- per Share on the 12th March, 1927.
5/- per Share on the 12th April, 1927.
5/- per Share on the 12th May, 1927.
Payment in full may be made upon Allotment.

DIRECTORS.

ARTHUR HENRY DILLON, J.P. (Chairman), Barton Lodge, Steeple, Aston, Oxford, Chairman of the Oxford Portland Cement Co., Ltd.
SIR JOHN BRUNNER, Bart., 43, Harrington-gardens, S.W.7. Director of Imperial Chemical Industries, Ltd., and Brunner, Mond and Co., Ltd.
LORD IVOR SPENCER CHURCHILL, 9, Chelsea Embankment, S.W.1.
GEORGE VINCENT MAXTED, M.Inst.C.E., Rhondda, Barnet-lane, Elstree, Herts. Director of the Dunstable Portland Cement Co., Ltd., &c.
HENRY MOND, 45, Green-street, Park-lane, W.1. Director of Imperial Chemical Industries, Ltd., and The Mond Nickel Co., Ltd.
HERBERT STEWART WILSON, of 13, Southwell-gardens, S.W. Barrister-at-Law.

BANKERS.

WESTMINSTER BANK, LTD., 4, Bartholomew-lane, E.C.2. Head Office, 41, Lothbury, E.C.2, and Branches.
ROYAL BANK OF SCOTLAND, 3, Bishopsgate, E.C.2. Head Office, 36, St. Andrew-square, Edinburgh, and Branches.

CONSULTING ENGINEERS.

MAXTED and KNOTT, LTD., 82, Victoria-street, S.W.1.

CONSULTING CHEMISTS:

WOODCOCK and MELLERSH, 76 and 78, Petty France, Westminster.

SOLICITORS:

HERBERT OPPENHEIMER, NATHAN and VANDYK, 1, Finsbury-square, E.C.2 (to the Company).
COX and CARDALE, Tower Royal, Cannon-street, E.C.4 (to the Vendors).
GRUNDY, KERSHAW, SAMSON and Co., 6, Austin Friars, E.C.2, and 31, Booth-street, Manchester.

AUDITORS.

WHINNEY, SMITH and WHINNEY, 4B, Fredericks-place, Old Jewry, E.C.2.

BROKERS.

SIMPSON, MILLER and SPRINGER, 5, Great Winchester-street, E.C.2, and Stock Exchange.
DAVID Q. HENRIQUES and Co., 13, Pall Mall, Manchester, and Stock Exchange.
E. A. and R. M. BAKER, Lloyds Bank Chambers, Baldwin-street, Bristol, and Stock Exchange.

SECRETARY AND REGISTERED OFFICE.

F. R. ALLEN, 3 and 4, Clement's Inn, Strand, W.C.2.

Figure 16; The Times newspaper advert for the share issue of the Oxford & Shipton Cement Ltd.
Courtesy C. Judge collection.

The directors were listed as:

Arthur Henry Dillon, J.P. – (Chairman) Barton Lodge, Steeple Aston, Oxford.
Chairman of the Oxford Portland Cement Co, Ltd.

Sir John Brunner, Bart – 43 Harrington Gardens, S.W.7.
Director of Imperial Chemical Industries Ltd., and Brunner, Mond and Co. Ltd.

Lord Ivor Spencer Churchill – 9 Chelsea Embankment, S.W.1.

George Vincent Maxted, M. Inst, C.E. – Rhondda, Barnet Lane, Elstree, Hertfordshire.
Director of the Dunstable Portland Cements Co., Ltd.

Henry Mond – 45, Green Street, Park Lane, W.1.
Director of Imperial Chemical Industries, Ltd. and the Mond Nickel Co., Ltd.

Herbert Stewart Wilson – Barrister-at-Law – 13 Southwell Gardens, S.W.

Other parties involved were:

Solicitors – Herbert Oppenheimer, Nathan and Vandyk – 1 Finsbury Square E.C.2.

Bankers – Westminster Bank and Royal Bank of Scotland.

Consulting Engineers – Maxted and Knott Ltd., 82 Victoria Street, S.W.1.
Consulting Chemists – Woodcock and Mellersh – 76 and 78, Petty France Westminster.

Results from the vendors' Consulting Engineers stated that they had visited the works at Kirtlington, the Sales depots at Oxford, Birmingham and Coventry, and had seen the land at Shipton that the new works were to be built on.

The Company Consulting Engineers then reported profits for the last six years from the Kirtlington works as:

31st of December,	1920	£13,526	12s	9d	
" "	1921	£4,949	12s	6d	(stoppage due to the General Coal Strike)
" "	1922	£9,335	13s	6d	
" "	1923	£11,541	12s	3d	
" "	1924	£5,774	18s	10d	(stoppage owing to Builders' Strike)
" "	1925	£8,823	0s	4d	

These profits over the six years averaged out at £8,991 0s 0d per year but the consulting engineers believe that if these new works at Shipton-on-Cherwell are carried out as planned, then the profits from the existing Kirtlington plant could be increased to £11,000 0s 0d.

However, due to the increasing general demand for Oxford portland cement and after careful examination of all the ageing properties and the exhausting natural materials at the Kirtlington works, they were convinced the best course to go forward by this new company was to construct a new modern works at Shipton-on-Cherwell on the freehold land already owned by OPCC, which was situated about two miles from the Kirtlington works. There would still be plenty of room for additional works if need be but with the two proposed large rotary kilns, the target figure of 125,000 tons of cement per year was achievable. Further reasons for building a new works would be the fact that the site contained the same raw materials as those found and used at Kirtlington; the works would have easy access to the Great Western Railway and so open sales to a vast new market and still have access to the Oxford Canal but, most of all, direct access to the national road system. The river Cherwell passes through the property and so a ready-made water supply was guaranteed. On site was an existing quarry called 'Bunkers Hill,' which would provide a deep and ready-made quarry face to work from. This had been inspected and proved that there was enough material for at least 70 years production and, stated from the consultant engineers, that there was also material suited to produce Tarred Macadam, Hydrated and Lump Lime, which was used in modern road making for which there was a large demand that was growing fast.

With the railway alongside and the new works obtaining its own sidings, the coal needed from the Midlands could now still come in by canal but also much quicker by train. The price at this present time, via the railway, averaged 16s per ton.

The provisional cost was estimated using all the modern improvements for economical and high-quality production, including rail sidings, internal roads, own electrical power plant, cement sacks, buildings to include office space and laboratory accommodation, all professional fees plus additional cost for employees housing (estimated at £10,000), giving an estimated total of £287,000.

The prospectus then states the value of Oxford Portland Cement Co. Ltd. for purchase by the new Oxford & Shipton Cement Ltd. would be: £108,250 0s 0d

This total valuation was made up by:

The Washford Pits (Kirtlington Cement works) site; several detached properties, the leasehold lands, buildings, machinery, transport, agricultural implements, office and laboratory equipment, stocks of coal and raw materials that are partially manufactured and finished, leasehold properties, plants, stock of cement at the Company's depots at Oxford, Birmingham, and Coventry. As at the 31st of December 1926:

Value £66,959 0s 0d

The main property at Shipton-on-Cherwell of about 216 acres, together with the detached area of about 5 acres with all the buildings and trees thereon, live and dead stock, acts of husbandry in the Parish of Kirtlington: Value £41,291 0s 0d

TOTAL £108,250 0s 0d

Interestingly this purchase price for the new company to buy was to be paid by a large cash sum, to the value of £58,250 and the remaining £50,000 in fully paid up £1 shares.

Then came the new company's estimated Capital Expenditure for the new works at Shipton of £500,000, which was broken down into:

> Erecting the new works and equipping it complete, including a power plant, railway sidings, roads, employees housing, sacks and architects' fees and expenses £287,000 0s 0d
>
> Preliminary expenses including advertising and printing, but exclusive of Government registration fees, underwriting and brokerage estimated at £18,000 0s 0d
>
> Government registration fees, underwriting and brokerage fees at £28,000 0s 0d
>
> Cash payable to the Vendor £58,250 0s 0d
>
> Working capital which the directors anticipate will carry out (if thought expedient) some of the extensions (in addition to the new cement factory). £108,750 0s 0d
>
> TOTAL £500,000 0s 0d

The allocation of shares was listed at the birth of the new Oxford & Shipton Cement Ltd. as: – Mr Dillon – 10,000: Sir John Brunner, Bart – 25,000; Lord Ivor Spencer Churchill – 15,000; Mr Maxted – 5000; Mr Mond – 15,000 and Mr Wilson – 1,000 shares.

It was stated to the new proposed company that the Kirtlington works were at present supplying various Government Departments, Local Corporations, Tramway and Canal companies and Housing schemes and so the new Oxford & Shipton Cement Ltd. was born.

The Oxford Portland Cement Co. Ltd. (although purchased by the Oxford & Shipton Cement Ltd.) was still in existence until March 1927, when it was voluntary wound up *(see Figure 18)*.

The first Annual General Meeting of the Oxford & Shipton Cement Limited was held in March 1928 and again Mr A.H. Dillon was in the chair. Extracts of his report follow:

> Gentlemen – You have all received the statement of accounts and the balance sheet for the year ending 31st of December 1927.
>
> The profit recorded is £4,297 11s 0d which is not so satisfactory as we had hoped. Owing to the coal strike, the cost of coal was particularly heavy in January and February and we did not enjoy cheap coal until the last three months of the year. Right up to December last, £31,377 18s 9d was expended on the new works at Shipton. You will see from the new works construction report, that good progress has been made with erection of the new works. Three lines of railway sidings have been laid by the GWR and almost completed, with room for a fourth if needed.
>
> One kiln is now in the course of being delivered by Edgar Allen & Co. Ltd. of Sheffield and the kiln piers and kiln bed are ready for the erection of the kilns. A new concrete road from the works to the Oxford main road is almost complete. Twenty-four houses have been erected

on behalf of the company by the Industrial Housing Association (No.2) Limited and several already are now occupied by the company's employees.

The Company has entered a contract with the Wessex Electricity Company for the supply of all electric power on very favourable terms. This means that there has been a saving of £40,000. This has enabled us to put up extra plant for making rapid hardening cement and to increase the output from 125,000 to 150,000 tons per year.

It is hoped that the plant coming from Edgar Allen & Co. Ltd.* will be completely erected and producing cement by the end of September next.

Note: A row of houses was constructed in 1926/27 in the typical 1920s design-style of the time, at a site called Bunkers Hill**. The houses numbers 1 and 2 were built for the cement works managers, as they were set back and angled from the others, that were numbered 3 to 22. Numbers 23 and 24 were originally the local post office and shop, but now have been converted to just normal houses.

The houses faced the Shipton quarry, except for the managers houses as they were angled differently. Adjoining, or near to the managers house was a concrete water tower on metal legs. This had been built as there was no water supply where the houses were built and the cement company found an underground spring nearby and pumped up the water by means of a windmill type pump. This was a problem when there was no wind as it resulted in no water, so an electric pump was fitted and ran until 1980, when finally, the water was piped from the village of Shipton-on-Cherwell. This caused a further problem as the pipe was too small, so the old water tower was brought back into service to ensure a constant supply of water. Evidence today of the post office can be seen where the post box still stands.

A social club was formed in 1935, and the clubhouse latterly consisted of two ex-RAF wooden huts and which were erected about 1947-48 and were in poor condition by the 1970s. The club complex was situated to the rear of the Bunker Hill houses. The club was registered for 150 members at any one time and consisted of the main hall, lounge, small kitchen, small cloak room, toilets and a bar/stock room. It was constructed purely for the cement company employees and their families. It included an outdoor 6-Rink Bowling Green, which was a first-class facility and was approved by the Oxfordshire Bowling Association for use in County Competitions. One football pitch, which was not used towards the final years, due to the age of the workforce. One cricket square that was a first-class area and used by local sides for league matches and had been passed by the Oxfordshire Cricket Association. Tennis courts, that again were in good condition and regularly used by the membership and it had other smaller sports facilities. However, there were no changing facilities or showering accommodation. The club at one time supported two darts teams of both genders – table tennis, dominoes, cards etc. There was also a television in later years, juke box, space invaders and fruit machines. Weekly bingo social evenings were arranged with discos and dance nights also organised – in all at its peak, a very well provided company social club.

Due to the decline of the cement works in later years, the social club was eventually sold off in 1992 to the then present social club steward, leaving just the bowling green in use. This too has now been

* In the Scotsman newspaper for Saturday November 12th of 1927, it was stated that the makers of crushing and grinding machinery for the lime and cement industry are extensively well booked and this week have made further important additions to their order books. One such contract is for a plant to produce 120,000 tons of cement a year. This has been placed by the Oxford & Shipton Cement (Ltd) with the Edgar Allen & Co. There will be two rotary kilns 200 feet long, with crushing, grinding, and packing equipment, and they will be fitted with turbo-pulverisers for pulverising the coal and firing. Two grinding mills will deal with the limestone and shale after it has been crushed by a giant hammer crusher. A further two grinding mills will be utilised for grinding the cement. Electric motors, totalling 4000 horse power will be installed for driving the machinery.

** The area called Bunkers Hill is listed in the National Monuments Records as Bunkers Hill Long Barrow and is recorded as 70m in length by 17m wide, 0.3m in height and its axis is 15 degrees west of true north and that it appears to have no side ditches. However, there is no evidence to suggest that it still exists. The area has been surveyed and walked the entire length and it appears that the barrow has disappeared, having been ploughed in over many years.

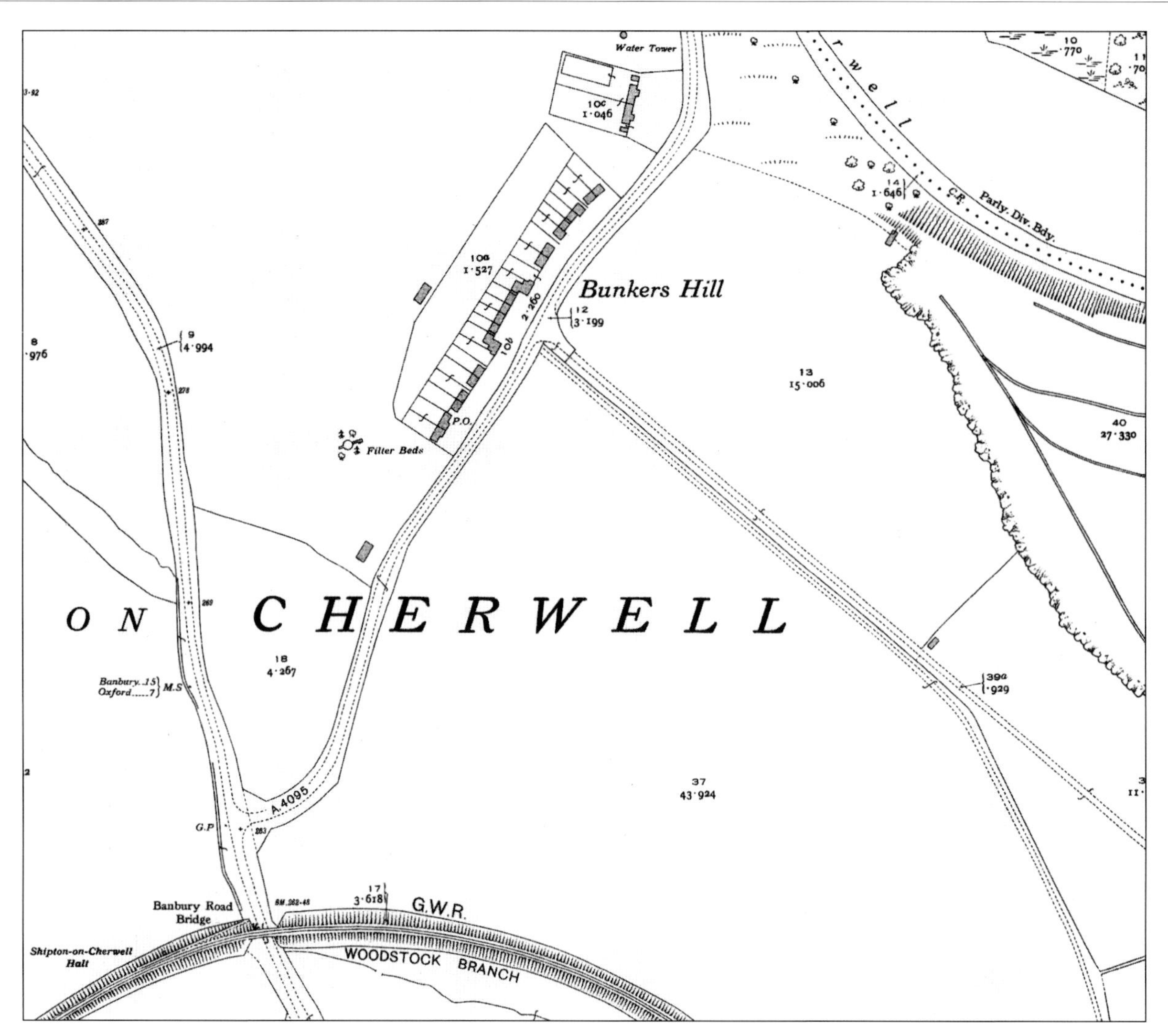

Figure 17: This map shows the relationship between the cement company's Bunkers Hill houses and the water tower (top), the Shipton-on-Cherwell cement works internal railway tracks (right) and the GWR Shipton-on-Cherwell Halt on the Woodstock Branch line (bottom left). Note the area behind the houses allocated for football and cricket pitches, bowling greens and tennis courts – lastly the social club building. The access road into the cement works starts right opposite the line of houses. At this 1930s date, the quarry face is relatively far away from the houses. This map also shows how the main road junction had become two 'T' junctions which had been constructed from the dangerous early four-way junction shown in Figure 15.

Reproduced courtesy National Library of Scotland under the Creative Commons Attribution licence.

Plate 34: A poor-quality image of the Oxford & Shipton Cement Ltd. Social Club, situated just behind the row of Bunkers Hill houses. This had been constructed from old RAF billet dormitory huts!

Courtesy of the 'Blue Circle' (the house magazine of Blue Circle Cement) Volume 10, No. 1 (January 1956).

Plate 35: An early 1930s photograph of the new houses built at Bunkers Hill. These were constructed for the employees of the new plant at Shipton-on-Cherwell cement works. Being built some months ahead of the new Shipton cement works, several Kirtlington cement works employees had already moved in, mainly those who were going to transfer to the new cement works site. Only the managers' houses were detached and built side-on to the road, all the rest were semi-detached in a straight line.
Courtesy Oxfordshire County Council, Oxford History Centre; ref POX0196206.

Plate 36: This small housing complex even had their own shop and post office situated in numbers 23 and 24. The post office officially opened on 2nd of January 1927 and the post box still exists today situated just out of view to the left of the photograph. *Courtesy Oxfordshire County Council, Oxford History Centre; ref POX0196213.*

sold, in 2009, due to lack of demand. The sports field has been the subject of several proposed housing developments but at present it is in use by the local garden centre.

The contractors called in to build the new Shipton-on-Cherwell cement works are reported to have been Edwards Construction Company Limited and they are said to have owned 2ft gauge tank locomotives, possibly Kerr Stuart 'Wren' class, plus some 'petrol locomotives'. This cannot be positively confirmed but was information reported by a visitor to the factory.

By February the 28th, 1929 the newspapers reported the second Annual General Meeting, held again at Westminster House, which was entitled "Completion of the Works".

Mr A.H. Dillon was again in the chair and said:

> The company has sustained a severe loss in the death of Sir John Brunner, whose experience and advice has been of very great value to the company from its inception and I am glad to say his son, Sir Felix Brunner has joined the board in his place.
>
> Also, Colonel E.P. Dillon, C.M.G., D.S.O., has also recently joined the board in the place of Lord Ivor Spencer Churchill, who retires by rotation and does not wish to be re-elected.
>
> I refer to the accounts, they show a loss of £10,954 3s 11d, but bringing the balance forward from January 1st 1928, the balance sheet then shows a net loss of £6,656 12s 11d. The loss of trading is the result of the price-cutting competition which has occurred throughout the greater part of this year. With most of the other outside cement manufacturers, this company joined the Cement Makers' Federation in October of last year, but after very careful consideration, notice has been given terminating this company's membership and on May 1st next, the company will have complete freedom in the sale of its products.
>
> To the end of December 1928, £264,057 9s 3d had been expended on the erection of the new works at Shipton-on-Cherwell. I had hoped to inform you today that the new works were complete and up and running, but despite every effort to achieve this end including the employment of a night shift all this summer, I have been disappointed. However, I have every reason to believe that within the next few weeks the factory will be complete, and will constitute one of the most up to date plants of its kind in the country.

The report goes on to say under the heading "Reaching the production Stage":

> From the time the raw material is dug from the quarry until the finished packed cement is automatically delivered into the truck, it will never once be touched by hand, and in consequence of these labour-saving devices and appliances, it is confidently anticipated that a works staff of less than 150 persons will be required to produce 150,000 tons of cement per annum – namely an average of 1000 tons per man per year. The raw material mills will shortly be in operation and it is expected that the manufacture of cement will be in full swing by the end of March, and that maximum production will rapidly be attained. We hope to produce the best quality cement at the rate of 150,000 per annum
>
> The company has increased its sales organisation and now has established a depot with a salesman, who has a motor lorry and driver in each of the following towns and cities – namely; – Oxford, Coventry, Birmingham (2), Newbury, Swindon, Exeter, Taunton, Bath, Bristol, Bournemouth, and Slough. Further depots are about to be opened in Reading, Northampton, Leicester, Nottingham, and Wolverhampton.
>
> It is not anticipated that the company will experience any serious difficulty in selling practically the whole of its output at remunerative prices.

Sir Felix Brunner seconded the resolution, which was carried unanimously and the meeting closed.

Chapter Four

The New OXFORD & SHIPTON CEMENT LIMITED Plant is Built and Operating.

Firstly, a plan of the internal railway in the new Shipton works seen in *Figure 18* on the 1937 OS map (published in 1939).

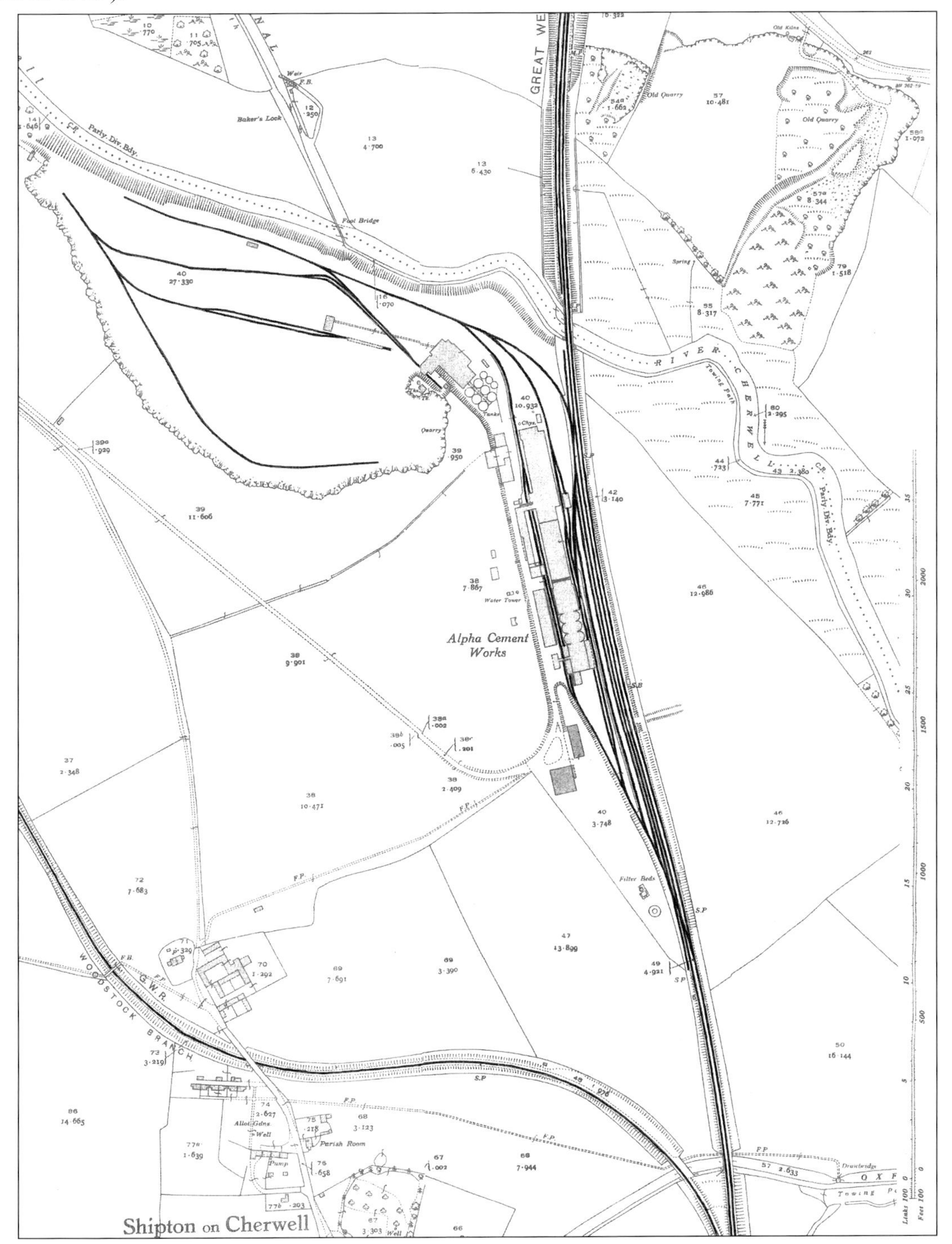

Figure 18: The OS map of the area showing the internal railway system of the Shipton-on-Cherwell cement works and the GWR main line from Oxford to Birmingham main line plus the GWR Woodstock branch line. This map refers to the post 1934 title; Alpha Cement Works.

Reproduced courtesy National Library of Scotland under the Creative Commons Attribution licence.

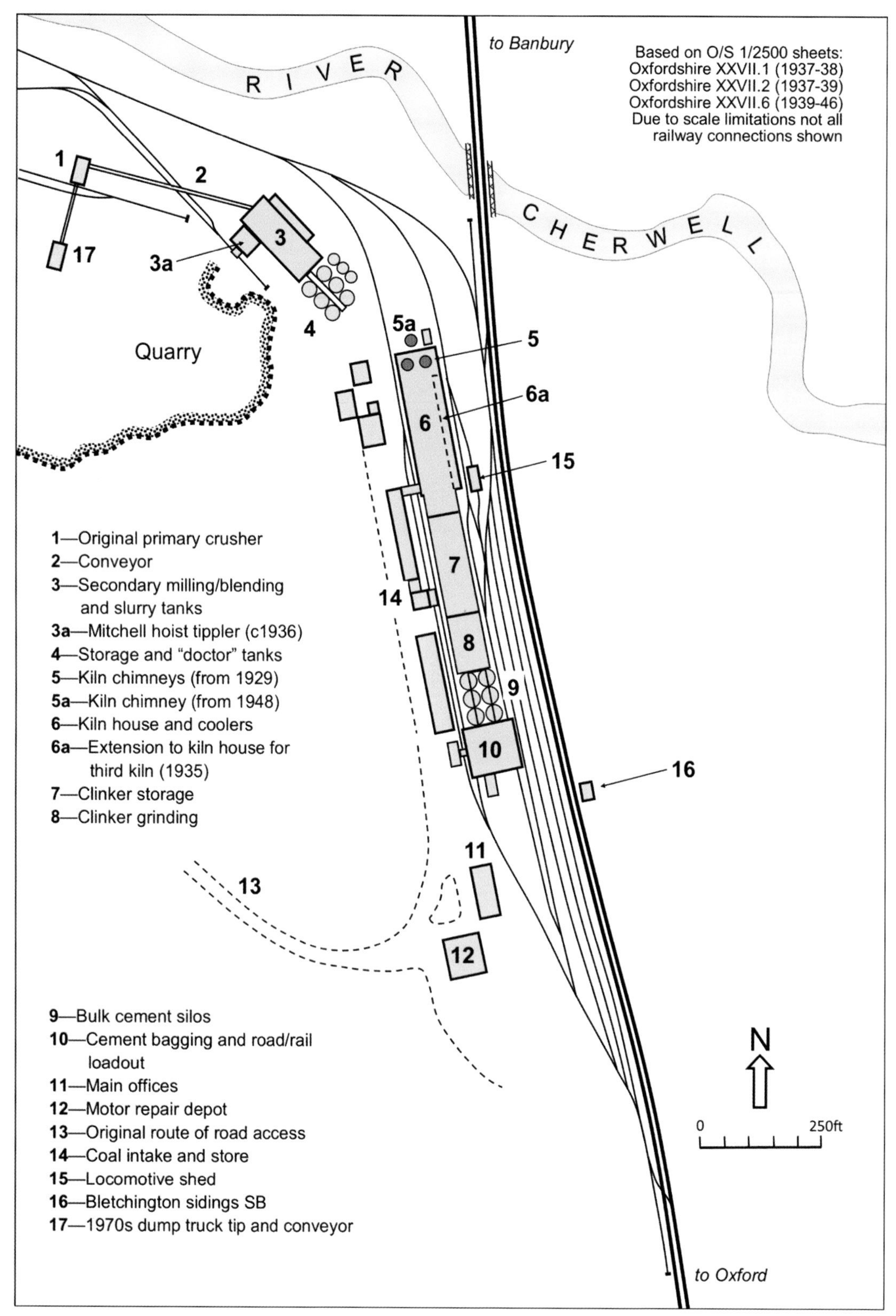

Figure 19: Shipton-on-Cherwell cement works, like many other factories, evolved over time to take account of improved processes and modern up-to-date equipment. This drawing is based on OS maps that were surveyed in the 1937 to 1939 period, when the Shipton works had already undergone several adaptations from when new; a third kiln had been added in 1933 and a larger crusher in 1936. The most marked visual change was with the kiln chimneys; originally, when built, the plant was equipped with two short chimneys (see Plate 39), one for each kiln, however, in 1948 those were replaced by the single 250 ft high structure (see Plate 55) that provided a very valuable navigation aid for Kidlington airfield operation as described in the Introduction. This new 1948 chimney is located on the above drawing (5a). Also shown is the location of the new tipping point, fed by the dump trucks which replaced rail use in the quarry in the early 1970s. *Drawing by Chris Down.*

Plate 37: The Shipton cement works in the early 1930s showing details of the quarry equipment – GWR wagons, Ruston 25 steam excavator and quarry internal rail lines. This view is almost certainly looking from Bunkers Hill. Courtesy Oxfordshire County Council, Oxford History Centre; ref POX0196209.

Plate 38: As no view of the Shipton quarry face on its own has come to light, this later view in 1970s shows the quarry face height compared to the locomotive No. 3 AB 2041 and wagons seen in the photograph. The various levels of the different quarry face materials are very visible. Courtesy Industrial Railway Society collection.

It is very difficult to say exactly when the first sod was turned and the first pouring of concrete made on the new works at Shipton-on-Cherwell but, from the first Annual General Meeting report, we know that the work of clearing the site had begun on the 21st of February 1927, also that the relatively small amount of £31,377 18s 0d had been spent. The approach road had been laid, some of the building's foundations prepared and the first kiln was due to be delivered from Messrs Edgar Allen & Co. Ltd. The remaining plant machinery had been ordered also from Edgar Allen & Co. Ltd.

When the plant was installed, it comprised of two-200 foot long 9-foot diameter rotary kilns which were fed with wet-ground material by three Stag Combination Tube Mills – themselves supplied by the Ergo crusher. This crusher reduced stone from blocks up to 4 ft square to about ¾ inch cubes in a single operation. The output was a staggering 120 tons per hour and it could handle and crush a piece of rock weighing up to one ton in 15 seconds!

The first works manager appointed, who oversaw the plant being built and equipped, was H.P. Lidgey.

The all-important railway connection to the Great Western main line railway was not in and the Bletchington Cement Siding signal box not yet built (it was constructed in November 1927), however, the three cement works sidings had been laid by the Great Western Railway.

The third Ordinary Annual General Meeting of the Company was again held at Westminster House, Old Broad Street, London E.C. on the 28th of February 1930, with A.H. Dillon again in the chair. He read out the following statements:

> Ladies and Gentlemen. The profit and loss accounts show a gross profit on the manufacturing account of £21,081 12s 3d and after making provision for rates, insurance, maintenance, carriage, salaries, expenses of the depots, motor cars etc. and allowing £9,234 3s 2d depreciation, there is a net loss for the year of £37, 930 3s 4d.
>
> The amount expended on the new Shipton works, including plant and machinery at cost amounts to £370,634 12s 9d and cars, lorries, depots, and equipment at cost stands at £13,982 12s 10d.
>
> The stock of cement, clinker etc. has been valued on a conservative basis at £36,12s 10d. The bank loan stands at £120,000.
>
> Those of you who have studied the balance sheet may realise that the cash position has given the directors some concern, but it is not anticipated that any further capital will be required.
>
> The company's sales organisation is now very well established and more depots have been opened and we now have them in the following towns and cities; – Bath, Bedford, Birmingham (2), Bournemouth, Bristol, Coventry, Gloucester, Leicester, Chelsea, Kingston, Sudbury, Westbourne Park, Ealing, Woolwich, Northampton, Nottingham, Oxford, Portsmouth, Reading, Salisbury, Sheffield, Slough, Southampton, Taunton, Walsall, Watford and Wolverhampton.
>
> The sales are progressing satisfactorily and the number of customers on our books increasing daily.
>
> As regards the results of this year's working, it is far too early for me to hazard a prophecy as it is quite uncertain what prices we may obtain over the whole year. The price of cement like that of all commodities is now at a very low level. There is no doubt that, however, that there is quite a future for the cement industry in this country. In due course the demand is bound to catch up with supply. When this happy situation arrives, there is no doubt that your company will be in a very favourable position to take full advantage of improved market conditions.
>
> To turn for a moment to the manufacturing side from the new works. The quarry is developing extremely well and the raw materials are providing excellent and evenly balanced and we find we can use the whole of the materials, except for a few inches of overburden.

> Apart from certain minor adjustments, the works are completely erected and are now running satisfactorily, and ordinary and rapid hardening cement of the highest quality are now being made. The company has received numerous reports from our customers as to the excellence of the company's product.

He finished with a statement that an interim progress report would be issued halfway through the year and the proceedings then ceased.

The fourth Ordinary General Meeting seems to have been postponed and not therefore recorded.

Plate 39: A fine aerial photograph of the Shipton plant as originally built taken in 1930. It clearly shows the new approach road running across the top of the photograph with main GWR main line from Oxford to Banbury running just in front of the works and across the canal bridge. The Oxford Canal snakes across the middle of the photograph and the surrounding area appears to have been in flood at the time. Note the two original small chimneys in the middle of the photograph. The small single building in the middle of the photograph is the locomotive shed, whilst far over to the left, the small Bletchington cement sidings signal box can be seen.

Courtesy Historic England; ref EPW 031375.

The fifth Ordinary General Meeting of the Oxford & Shipton Cement Ltd. was held on the 29th of February 1932 as usual at the Winchester House, London with Mr A.H. Dillon as Chairman.

This year's working has resulted in a reduction of nearly £20,000 in the amount owed by the company. The item freehold land and houses at Shipton is somewhat higher, due mainly to the payment of part of the purchase price of about 60 acres of land in the parish of Shipton, containing a large deposit of raw material for cement manufacture which we were able to obtain at a satisfactory price.

Improvements to the works
Rather more than £6000 was spent during the year in various improvements in the works. This includes the provision of an extra locomotive and a steam digger in the quarry. Two new additional elevators of a large capacity have been installed for the conveyance of clinker and cement. The spare engine, new excavator digger and elevators has given the opportunity of effecting repairs without stopping production. I may say that your directors are very well satisfied with the results of the expenditure on the capital account during the year.

At the old Kirtlington quarry a new stone-crushing plant is in course of erection and it is hoped and expected that the sale of crushed stone for road purposes will be a source of small additional profit to the company. The only item which calls for comment on the debit side of the profits and loss account is "Administration, selling and general expenses" £37,882 which is £7,100 less than the previous year, chiefly owing to economics in the sales department.

During 1931 the production and sales of cement were slightly higher than during 1930. Perhaps I should tell you that proceedings have been launched recently by the owners and occupiers of certain lands in the neighbourhood of the company works at Shipton-on-Cherwell in respect of emissions from the company chimneys, which are alleged to be noxious. It would be manifestly improper for me to comment upon these proceedings while the matters complained of are *sub judice*, but I believe I am entitled to say that when action was threatened, your directors sought the opinion of an expert who advised that no nuisance of any kind had been committed and no actionable damage had been caused by the company.

A year of Intense Competition
This year of 1931 will be remembered in the cement industry for the intense competition for orders, which ended in a scramble in which competing companies quoted for contracts at any price without reference to costs of production. This ended by one of the large combinations of companies going into liquidation in the summer. After this the trade met and agreed to stop price warfare, since when our average monthly sale price has risen steadily, but is still below that which we were obtaining at this time last year. When the contracts which were made at cut prices are worked out, all cement manufacturers ought to receive an adequate return for their products.

Although some public and private work has been stopped or suspended, I still think that the consumption of cement will be nearly, if not quite as much in 1932 as it was in 1931 and now as less cement is being imported, due to us going off the gold standard, and more being exported for the same reason. I hope manufacturers will be able to keep their plants as fully occupied as before. Our sales for January and February have been slightly better than for the same months last year, and it is hoped that this may continue and that the company will reap the full benefit from the added machinery and of the changes the directors have made, which have improved the production of the works.

The report was seconded by Sir Felix Brunner and he paid tribute to the work done by the company's staff and directors during the year.

In letters held in the Oxford University New College archives, there are two dated 19th September 1931 from the solicitors of the college to the Oxford & Shipton Limited selling more freehold land at Shipton, consisting of 61.112 acres for the purchase price of £900. This is likely to be the agricultural land between the original purchase area shown in *Figure 15 (page 45)* and the GWR Woodstock branch line, into which the quarry expanded post-WW2.

We continue with the next year's Sixth Annual report on the 1st of March 1933 with the now Viscount Dillon again in the chair reading the report:

> During the year the bank loan has been reduced by £32,000 and now stands at £85,000. Sundry creditors stand at £21,033 against £16,891. Sundry debtors are £64,111, as against £48,847. The increases in these figures are of course mainly due to the increased production and sales. In previous balance sheets, the stock in trade and stores and spare parts, have appeared under one item but this time have been separated, and stocks of cement, clinker and raw material stand at £13,927.
>
> You will have noticed that the Kirtlington works have now been abandoned, as your directors are of the opinion that under modern conditions it is impossible that they could ever again be worked at a profit. When, therefore, the state of the market makes it necessary that we should increase our productive capacity to any large extent it will be the policy of the Board to make additions to the new Shipton works, rather than start the old works at Kirtlington. This brings me to the question of the state of the balance sheet in general, and I am asked by my four colleagues to say that in their opinions the time has come for considering the question of what may be described as the "cleaning up" of the balance sheet. This would involve the writing off a sum not exceeding 5s on the nominal value of each share and would enable the Kirtlington property to be written down to its real value and would also eliminate the heavy item of £48,205 for preliminary expenses. This cannot be affected without the consent of the shareholders, and the question of a decision on this point does not arise immediately, I wish, however, to say that I totally disagree with my colleagues on this matter.
>
> **Profit for the year**
> Regarding the profit and loss account, the gross profit for the year is £102,029 against £75,463 for 1931. The net profit is £31,565 as against £15,508 for 1931, and after deducting £32,305 standing to the debit of the profit and loss account for 1931, there is an available profit of £9,263, out of which your directors recommend a dividend of 1¾ %, less tax, requiring £5,062 and leaving £4,201 to carry forward.
>
> At present the works are in excellent running order, and for this we are greatly indebted to our works manager, Mr C.E. Dunleavie and his staff. In a year of extreme depression and keen competition the increased sales reflect great credit upon our commercial manager, Mr George Adams, and the staff under him. Speaking quite generally, the company has an excellent workforce composed of keen and energetic men whose loyal efforts we all greatly appreciate.
>
> Although the competition for orders was keen, the average price realised in 1932 was slightly higher than that for 1931. There was, however, a slow but steady fall in prices during the latter part of the year, and this tendency unfortunately continues. In view of this fact, it is extremely difficult to make any forecast of the probable results for 1933, but up to date both the production and sale of cement show a satisfactory improvement upon 1932.
>
> If some of the works and road construction and other schemes which were abandoned or postponed in 1932 are put into operation, your company is in an excellent position to take full advantage.

The Chairman then moved the adoption of the accounts and report. The motion was seconded by Sir Felix J.M. Brunner and after some heated discussion, was carried.

A resolution to appoint four additional directors was withdrawn by the Chairman and finally a vote of confidence of the Chairman and directors, closed the proceedings.

It is interesting to note that Albert Younglove Gowan was appointed Managing Director in 1934 of the Oxford & Shipton Cement Ltd., before the change to Alpha Cement Ltd. However, when Alpha Cement Ltd. was eventually acquired by APCM, he became a director of APCM and that continued until 1947.

It was during 1933 that a third kiln was added with a calcinator, that helped production to increase. It is interesting to note that the term 'calcinator', was a misnomer as nothing was calcined; it refers to using waste heat from the kiln to dry (but only partially) the wet kiln slurry feed before it entered the kiln, achieving a minor economy at the expense of operational complications. Shipton works was the first UK location to use this technology, which never became widespread.

It probably is now the best time to introduce a simple flow chart version of the two methods of cement production – the DRY process and the WET process. Both methods have been used by the two works covered in this book, although the dry process illustrated here (as proposed for the new factory at Shipton planned in 1982) was far more complex than the simple and basic wet process employed at the Kirtlington works.

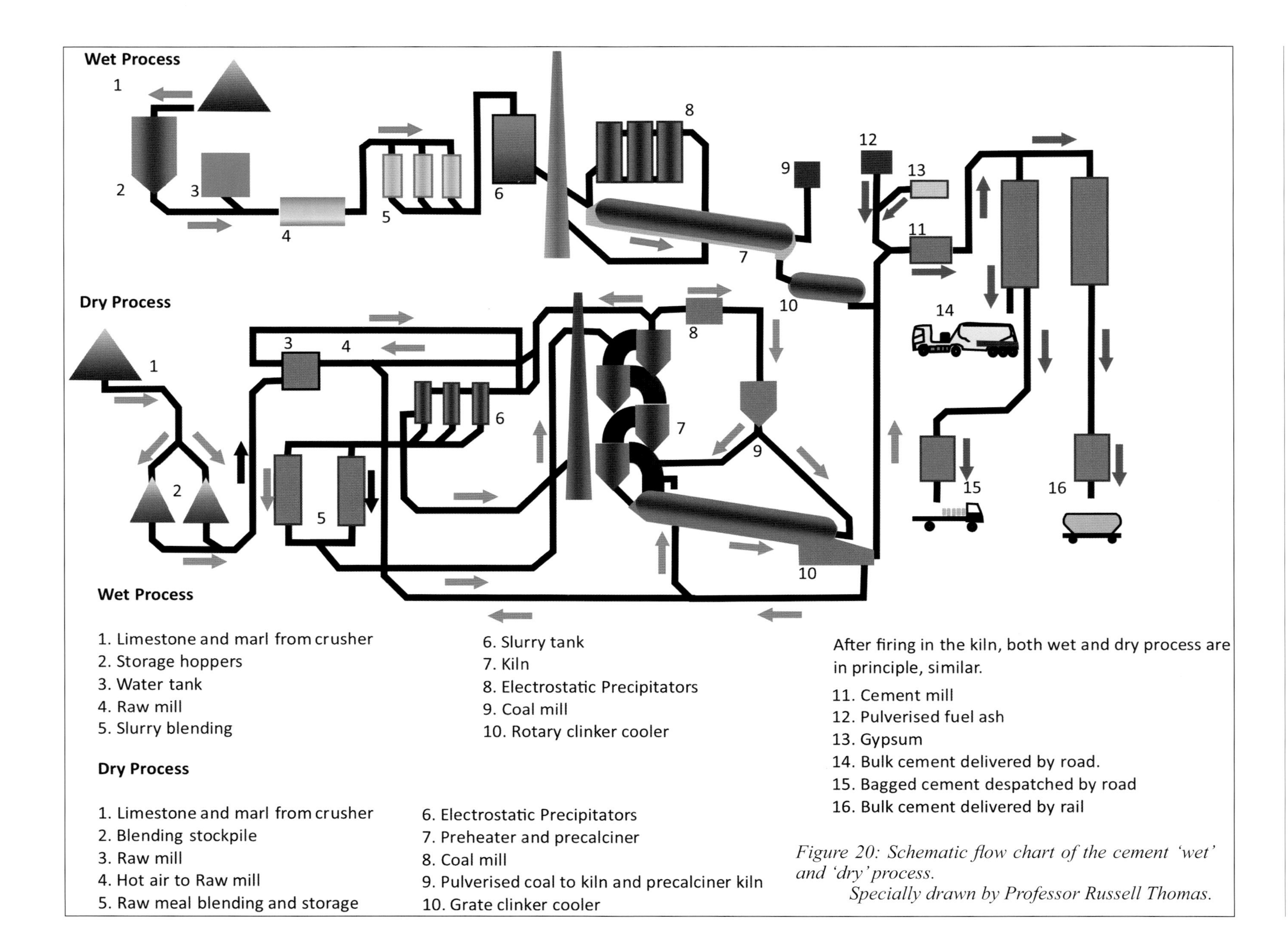

Figure 20: Schematic flow chart of the cement 'wet' and 'dry' process.
Specially drawn by Professor Russell Thomas.

The Shipton quarry had two levels that were called 'benches' *(see Figure 13)*. The upper bench had a face height of approximately 14 metres and incorporated three different clay beds. The lower bench incorporated the various stones and had a face height of approximate 6½ metres. In the early years the top soil was dug away by hand using shovels – a highly dangerous job!

The upper bench material was blasted and then dug out using a Ruston steam excavator (later an electrically powered navvy) and then loaded into the internal user wagons, propelled by one of the locomotives to the Mitchell wagon lift *(Plates 42 and 43)* and unloaded into a hopper feeding the crusher and thence via a 400-foot-long conveyer into storage silos.

Lower bench material was dug out by mechanical shovels and loaded into wagons, propelled to the Mitchell wagon lift by a locomotive and fed into the crusher intake.

In later years, after use of the quarry railway finished, quarried materials were hauled from the hydraulic face shovel to the crusher in 28-ton Aveling Barford Centaur dump trucks.

All the other ingredients came into the plant via the GWR main line and the coal wagons unloaded by the side tippler *(see Plate 41)*, The gypsum etc wagons were mainly unloaded by hand and shovel.

Plate 40: The reception hopper and feeder to the original 'Ergo' primary crusher installed in 1930, into which the dugout stone was tipped. Edgar Allen & Co. Ltd. of Sheffield was a prominent supplier of the first items of equipment to the Oxford & Shipton Cement Ltd. This facility was inadequate after 1935 when the original kilns had been uprated and a third kiln had been added, so it was supplemented by an impressive Mitchell hoist wagon tippler as seen in Plates 42 and 43. Courtesy Oxfordshire County Council, Oxford History Centre; ref POX0100223.

The crushed stone that came from the Miag Titan Hammer Mill would have started from the quarry face at about one metre size and then would be reduced to 25mm size. The crusher produced enough feed material for all three kilns.

The resultant stone was housed in a raw material store awaiting use and there was provision for outside storage on an uncovered stockpile, but this was seldom used due to low level dust problems. There were also difficulties in handling wet crushed materials which had been exposed to the elements. About seven day's consumption was stored here. It was then passed on to the wet raw milling machine, where the raw materials were mixed with water and slurry conditioning additives. This slurry was then blended to about 27% moisture content and stored in slurry tanks until needed and to allow time to mix properly.

The original kiln, supplied by Edgar Allen & Co. Ltd. in1928-29, was 200 ft long x 9 ft in diameter, the third kiln (1933) was the same length, but of a larger diameter.

Being called the 'wet process,' the raw material dries, then calcinated and then the clinker was sintered solely in the kiln tube and then cooled.

Coal, arriving at the works by rail was unloaded by a single 30-ton capacity tipper, then transported via a conveyor and elevator to coal bunkers before being fed by a conveyor into a hopper that was situated prior to the coal mill. This produced fine milled coal which was then fired directly into the kiln. The hot gases were drawn off the kiln hood for the purposes of drying the coal in the mill.

Plate 41: For incoming non-tipping main line wagons, a wagon tippler was installed at the coal reception hopper. This view in the 1930s shows that at least one coal supplier wagon was Spencer, Abbott & Co. Ltd., a large national coal merchant business, who had connections with the nearby Warwickshire Coalfield.

Courtesy Oxfordshire County Council, Oxford History Centre; ref POX0100227.

Plate 42: This fine view of the Mitchell tippler wagon lift showing details of the internal standard gauge stone wagons that were used on the Shipton plant railway. Note the two huge counter balance weights on either side of the tower. The crushed stone was transported by an inclined belt conveyor to storage which can be seen in the right. This operation must have been exciting to see in operation! Courtesy Industrial Railway Society.

Plate 43: An additional view of the Mitchell wagon lift showing its enormous lift height and its position within the plant. In the foreground is a heap of locomotive coal, so that locomotives bunkers could be topped up on the spot while waiting for the individual wagons to be tipped. *Courtesy Industrial Railway Society.*

The clinker produced by the kilns was then stored in a covered store with a capacity of 11,000 tons until needed. Gypsum arriving by rail from outside suppliers was stored in the same area.

The resultant cement was stored in six large 2000-ton silos and then was transported to the bagging area where it was bagged into various sizes of bags. In the early days this would have left the works by both road and rail but later mainly by road and bulk cement in road tankers, loaded via a chute.

In the early days of the works at Shipton nearly all the machinery had independent controls; later the works had a centralised control room!

In 1975 No's 1 and 3 kilns were shut down and a new Atritor Coal Mill was introduced to fire the remaining No. 2 kiln. When Blue Circle introduced their 'Packbind' cement, producing the clinker at the Barnstone works, it was Shipton which they selected to grind it and this continued until the works closed in 1986.

Back to the seventh Ordinary Annual General Meeting of the company that was held on February the 22nd, 1934 with Viscount Dillon presiding in the chair.

The chairman then reported:

> The bank loan has been reduced by £40,000 which then stood at 31st of December 1933 – £45,000. Sundry creditors were £22,050 against £21,033 in 1932. *(There followed considerable discussion on finances, which is not relevant here).*
>
> Motor cars, lorries etc stand at £14,371 as against £9,844 for last year being due to new motor lorries being purchased. The total profit available is £38,218 and your directors recommend a dividend of 4%.
>
> The Kirtlington works value is to be written down by £15,000 reducing the balance-sheet value from £45,962 to £30,462 which leaves a balance of £9,218 to carry forward to the 1934 financial year.
>
> These results have been obtained despite the fall in prices by manufacturing and selling more cement. Up to date this year we have made more in the same period of last year.
>
> **New Unit to Manufacturing**
> When we get our new third kiln running in April, we hope to be manufacturing about 50 per cent more cement. Since the issue of the report there are signs of an improvement in price, but it is too early to say whether a permanent improvement is likely.
>
> There are many reasons for anticipating an increase in consumption of cement this year, chiefly plans for slum clearance, water supplies and continued house building. I think, therefore, we can look forward with confidence to the trade in 1934.

(The remainder of the report was carried out by Sir Felix J.M. Brunner Bt. (Managing Director) reporting on the staff, the prices, and overseas markets).

The report was voted on and carried and the Chairman announced that the dividend warrants would be posted out on March 7th.

Opposite page:
Plate 44: This photograph shows the three 750hp motors driving the 'Stag Combination Tube Mills' which reduce the primary crushed stone down to ¾ inch size at the rate of 120 tons an hour.
Courtesy Oxfordshire County Council, Oxford History Centre; ref POX0100228.

Plate 45: The enormous powerful 'Centra-drive' and the discharge end trunnion of one of the Stag Combination Tube Mills.
Courtesy Oxfordshire County Council, Oxford History Centre; ref POX0100231.

EDGAR ALLEN
& CO LTD
SHEFFIELD

EDGAR ALLEN
& Co Ld
ENGINEERS

One interesting newspaper report was not mentioned at the AGM but had been reported in the Leamington Spa Courier on Friday 31st of August 1934.

A lorry fire.

Some excitement was caused in Oxford Street, London on Monday afternoon when flames were seen bursting from the body of a large steam wagon. The driver's attention was attracted and he pulled up opposite the Blue Ping Garage, which fortunately houses part of the Southern Fire Brigade apparatus. With the aid of a chemical extinguisher and buckets of water the fire was extinguished. The wagon was laden with cement in paper bags from the Oxford & Shipton Cement Works and the cause of the fire was the ignition of the bags by hot cement. Several ordinary sacks and ropes and one side of the wagon was damaged by the fire. The steam wagon was owned by the Dudley and Blowers Green Transport Company of Dudley.

Plate 47: One of the steam wagons owned by Dudley and Blowers Green Transport Company. It is possible this was the type used to transport cement in bags as these were heavy loads.
Authors collection acquired from an international auction site and reproduced under the Creative Common licence.

Opposite: Plate 46: The newly installed bagging plant seen here in the 1930s.
Courtesy Oxfordshire County Council, Oxford History Centre; ref POX0100272.

Plate 48: This wonderful 1930s photograph taken from the Enslow/Bletchington road bridge, shows the repositioned GWR Bletchington Station signal box on the right, situated north of the Shipton-on-Cherwell cement plant (in the distance) which had only recently been opened. The two small chimneys for the original kilns can be seen in the distant background, proving the date of the image and the complex GWR track work is very evident. Sometimes this signal box was used in conjunction with the cement works operation.

Courtesy Laurence Waters; Great Western Trust collection, Didcot.

Chapter Five

The GREAT WESTERN RAILWAY'S connection and the SHIPTON-on-CHERWELL CEMENT WORKS Internal Railway.

At this point, we should include a synopsis of the GWR Shipton-on-Cherwell halt, which played a small but important part in the history of the Shipton-on-Cherwell cement industry in Oxfordshire.

Shipton-on-Cherwell halt on the Great Western Railway's short Woodstock branch line (which opened on the 19th of May 1890) is very relevant to the Shipton cement works and is often forgotten. The halt was constructed in 1929, opened on the 1st of April 1929 to serve Shipton-on-Cherwell village, Bunkers Hill residents and the employees of the Oxford & Shipton Cement Company. It was one of 26 halts that the GWR introduced around the same time to accommodate the new steam rail motors, which it had just introduced to use on the local services. The new Shipton cement works in 1929 was also just starting production and had constructed 23 houses at nearby Bunkers Hill for some of its employees, mainly ones that had chosen to move from the original Oxford Portland Cement Co. Ltd. works at Kirtlington. However, this housing was completely on its own and without most amenities other than a small shop and post office.

The new halt allowed the cement works employees to now arrive by train from as far as Banbury, Oxford, Kidlington etc and Shipton villagers and Bunkers Hill residents could also venture out to Woodstock, Oxford and further afield, so the halt was a great benefit to all persons with their coming and goings!

Shipton-on-Cherwell halt was situated just 80 yards west of the main A423 Oxford to Banbury Road, next to the Banbury Road Bridge, on the north side of the branch line track, with the public access up a cinder pathway from the road, through the traditional kissing gate that was used on many GWR halts!

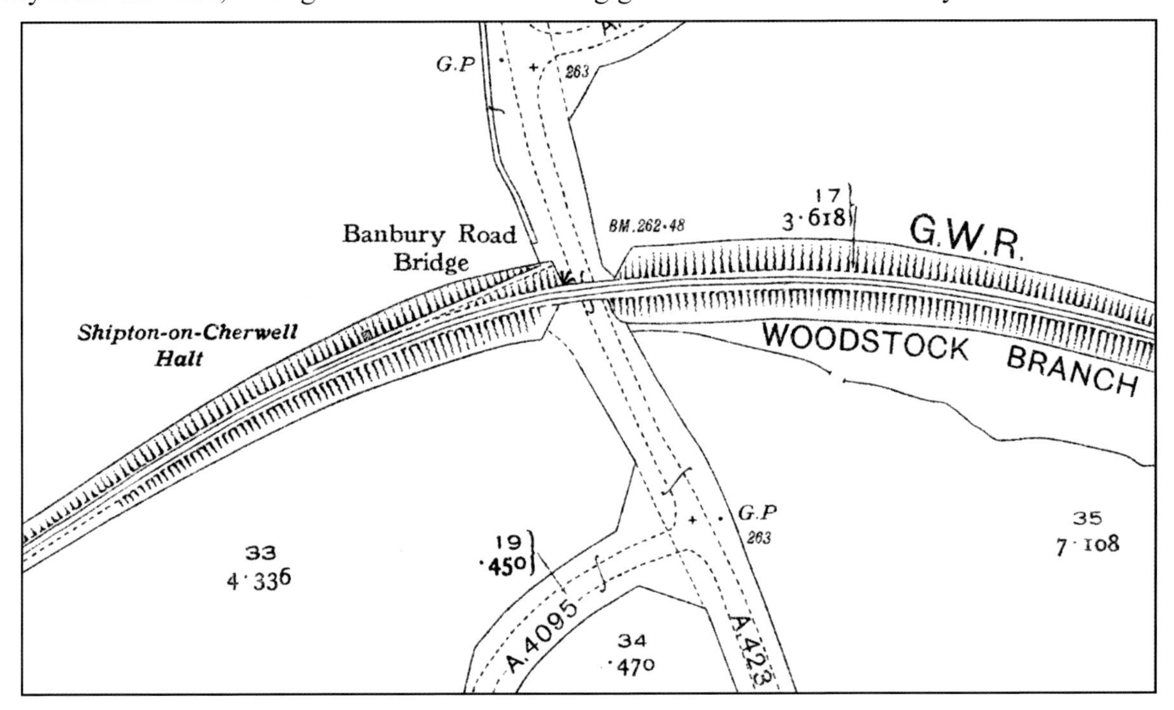

Figure 21: Small section of the 1939 25 inch Ordnance Survey map for Oxfordshire showing the halt and the changed road junction that at one time was a four-way junction, later, for safety, changed to two T junctions, approximately some 200 yards apart, as shown here.

Reproduced courtesy National Library of Scotland under the Creative Commons licence.

Plate 49: The original low, sleeper-built platform constructed for the GWR steam railmotors had disappeared by the time this 1940s photograph was taken. Note the Shipton cement works is visible on the right-hand side of the photograph, showing just how near it was to the halt.

Author's collection; courtesy of the late J.H. Russell collection.

Plate 50: The huge, bold nameboard on the railway embankment, situated on the side of the A423, advertising the importance of the halt. Note the entrance gate to the cinder pathway up to the platform with a typical GWR lantern style light plus the customary GWR trespass notice.

Author's collection; courtesy of the late J.H. Russell collection.

The halt was constructed using railway sleepers forming the very low platform needed to accommodate the newly introduced GWR railmotors on these branch lines. The rest of the halt consisted of a small draughty lean-to style wooden shelter for the passengers and two lanterns (which lit the halt, albeit dimly) that were mounted on stout wooden posts which were fitted with ornate finials.

The initial cost to the GWR was £160 and a further cost in 1933 of £120, when the height of the platform was raised to the standard passenger carriage platform height.

A large and impressive sign advertising the halt was erected on the side of the A423, alongside the girder bridge. This is shown in *Plate 50.*

Opening the halt seems to have made a significant change to passenger numbers for the Woodstock branch line; in the 1920s these were around17,000 per annum but, in the 1930s, they rose to 22,000.

However, after nationalisation in 1947, passenger numbers had fallen to below 9,000, with eight trains each way, some trains carrying less than ten passengers and some even empty!

8 Kidlington and Blenheim & Woodstock.—G.W.R.

ONE CLASS ONLY. WEEK DAYS ONLY.

	am	am	am	pm	pm	pm	pm	pm	pm		
Kidlington ....dep.	7 18	8 25	9 52	1225	2 36	4 3	4 57	646	937	......	..
Blenheim & Woodstock arr.	7 26	8 35	10 0	1233	2 46	4 11	5 5	654	945	...	..

8 Blenheim and Woodstock—G.W.R.

ONE CLASS ONLY. WEEK DAYS ONLY.

	am	am		am	pm	pm	pm	pm	pm	pm
Blenheim & Woodstockdep.	7 48	9 10	......	1112	1 5	3 25	4 35	6 13	8 17	10 0
Kidlington 9 ....arr.	7 56	9 18	..	1120	1 13	3 35	4 43	6 23	8 25	10 8

Figure 22: The 1926 GWR timetable for the Woodstock Branch, from the Bacons Guide, prior to the Shipton-on-Cherwell Halt being built. Courtesy Laurence Waters; Great Western Trust collection, Didcot.

GWR timetable 1931

BLENHEIM AND WOODSTOCK BRANCH

AUTO CAR—ONE CLASS ONLY.

No Block Telegraph on this Line. Form of Staff, round: Colour Red.

Single Line between Blenheim and Woodstock and Kidlington, worked by Train Staff and only one engine in steam at a time, or two coupled together.

DOWN TRAINS—WEEK DAYS ONLY.

Distance M. C.	STATIONS	Ruling Gradient 1 in	B Oxf'd Auto	B Auto Mixed	B Auto Pass.	K Goods	B Auto Pass.	B Oxf'd Auto	B Auto Pass.	B Auto Pass.	G Eng'e & Van	B Auto Pass.	B Auto Pass.
			A.M.	A.M.	A.M.	A.M.	P.M.	P.M.	P.M.	P.M.	P.M.	P.M.	P.M.
—	Kidlington dep	—	7 23	8 28	9 52 ‡	11 10	12 38	2 36	4§10	4 53	5 50	6 46	9 37
2 9	Shipton-on-Cherwell Halt dep.	60 R	7 28	8 34	9 57 ‡	—	12 43	2 41	4 15	4 58	— RH	6 51	9 42
3 57	Blenheim & Woodstock arr.	129 R	7 31	8 38	10 0	11 20	12 46	2 44	4 18	5 1	5 58	6 54	9 45

UP TRAINS—WEEK DAYS ONLY.

STATIONS.	Ruling Gradient 1 in	B Oxf'd Auto	B Auto Pass.	G Eng'e & Van	B Auto Mixed	B Oxf'd Auto	B Auto Mixed	B Auto Pass.	K Goods RR	B Auto Mixed	B Auto Pass.	B Oxf'd Auto
		A.M.	A.M.	A.M.	A.M.	P.M.	P.M.	P.M.	P.M.	P.M.	P.M.	P.M.
Blenheim and Woodstock dep.	—	7 47	9 10	10 30	11 42	1 0	¶ 3 25	4 35	5 20	6 12	8 17	10 0
Shipton-on-Cherwell Halt dep.	129 F	7 50	9 13	—	11 45	1 3	3 28	4 38	—	6 16	8 20	10 3
Kidlington arr.	60 F	7 55	9 18	10 38	11 50	1 8	3 33	4 43	5 30	6 22	8 25	10 8

‡ May run as a Mixed trip when necessary at point to point times shewn for other Mixed trips. When this arrangement operates 10.30 a.m. Blenheim and 11.10 a.m. Kidlington will not run. § Starts from Oxford on Sats. at 4.0 p.m., Kidlington arr. 4.10, dep. 4.11 and thence one minute later. ¶ Not a mixed trip on Sats., and works through to Oxford. Kidlington arr. 3.34, dep. 3.35, Oxford arr. 3.45 p.m.

Figure 23: The GWR working timetable for the Woodstock branch for 1931. Author's collection.

The last nail in the coffin fell on the 27th of February 1954 for the Shipton-on-Cherwell halt, when the last train ran with the halt officially closed on 1st of March 1954 and the track lifted in the first month of 1958.

Little, or almost nothing, remains of this structure except some wooden and steel rods that held the halt on the steep embankment and the worn edge of the platform peeping out from the undergrowth. However, small sections of the very overgrown branch line are still walkable today.

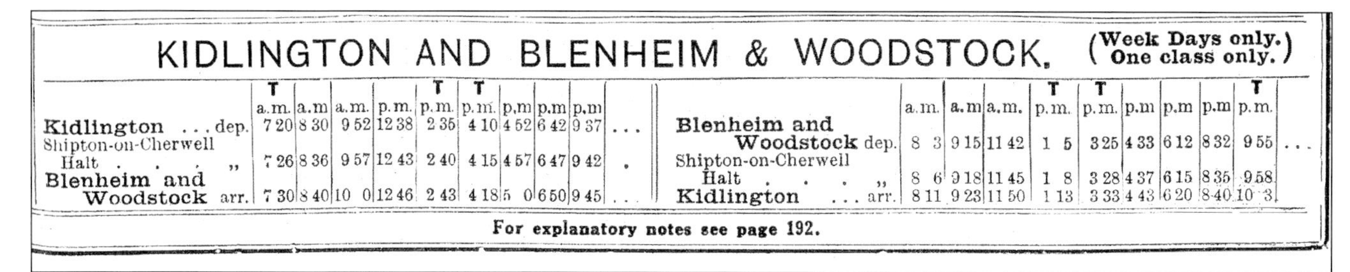

KIDLINGTON AND BLENHEIM & WOODSTOCK. (Week Days only. One class only.)

	T				T	T				
	a.m.	a.m	a.m.	p.m.	p.m.	p.m.	p.m	p.m	p.m	
Kidlington ... dep.	7 20	8 30	9 52	12 38	2 35	4 10	4 52	6 42	9 37	...
Shipton-on-Cherwell Halt . . . . ,,	7 26	8 36	9 57	12 43	2 40	4 15	4 57	6 47	9 42	.
Blenheim and Woodstock arr.	7 30	8 40	10 0	12 46	2 43	4 18	5 0	6 50	9 45	...

				T	T				T	
	a.m.	a.m	a.m.	p.m.	p.m.	p.m	p.m	p.m	p.m.	
Blenheim and Woodstock dep.	8 3	9 15	11 42	1 5	3 25	4 33	6 12	8 32	9 55	...
Shipton-on-Cherwell Halt . . . . ,,	8 6	9 18	11 45	1 8	3 28	4 37	6 15	8 35	9 58	
Kidlington ... arr.	8 11	9 23	11 50	1 13	3 33	4 43	6 20	8 40	10 3	

For explanatory notes see page 192.

D—On Saturdays, August 12th, 19th and 26th, conveys Through Carriages from Barry Island.
G—Saturdays excepted. **M**—One class only.
P—Passengers for the connecting train from Dudley to Birmingham via Old Hill, change at Blowers Green.
Q—On Saturdays until Sept. 2nd, inclusive, arr. Manchester (Victoria) 2.28 p.m. On Sept. 9th, 16th and 23rd, arr. Manchester (London Road) 2.36 p.m.
R—Via Shrewsbury and Wem. On Saturdays arr. 5.50 p.m.
S—Saturdays only.
T—Through Service between Oxford and Blenheim and Woodstock.
V—On Saturdays arr. 8.20 p.m.
X—One class only (limited accommodation).
Z—Saturdays only, and runs July 22nd to Sept. 2nd (inclusive) only.
a—On Saturdays arrive Chester 1.29, Birkenhead 1.57 and Liverpool (Landing Stage) 2.17 p.m.
b—On Saturdays arrive 5.30 p.m. **c**—Via Hartlebury.
d—Manchester (Exchange), via Chester. **e**—Restaunt Car to Oxford.
f—Manchester (Exchange) via Chester. On Saturdays until Sept. 9th (except August 5th) arr. 7.12 p.m. On August 5th, Sept. 16th and 23rd arr. 7.58 p.m. **g**—Reading West Station.
h—On Saturdays arrive Chester 1.45, Birkenhead 2.24, and Liverpool (Landing Stage) 2.37 p.m.
‡—Via Hartlebury. On Saturdays depart 2.12 p.m. via Kidderminster.
†—Five minutes later on Saturdays. §—Via Shrewsbury and Wem.
¶—Eight minutes later on Saturdays.

Figure 24: The working timetable for the Woodstock branch 1939.
Courtesy Laurence Waters; Great Western Trust Collection Didcot.

KIDLINGTON and BLENHEIM & WOODSTOCK. (ONE CLASS ONLY. AUTO CAR—) Week Days only.

SINGLE LINE between Blenheim and Woodstock and Kidlington, worked by Train Staff and only one Engine in steam at a time, or two coupled together. No Block Telegraph on this Line. Form of Staff, Round; Colour, Red

Down Trains.

Distance M.	C.	STATIONS.	Ruling Gradient 1 in	B Oxford Auto.		B Auto Mixed.		B Auto.	K Freight RR SX		B Auto. ‡	B Auto K SO	B Oxford Auto.		B Auto.		G Engine and Van. K	B Auto. MSX	B Auto. SO	B Auto. MO	B Auto. N	B Auto. V SO
				a.m.		a.m.		a.m.	noon.		p.m.	p.m.	p.m.		p.m.		p.m.	p.m.	p.m.	p.m.	p.m.	p.m.
—	—	Kidlington .... dep.	—	7 23	....	8 40	....	11 15	12 0	....	12 38	1 10	3 0	....	4 10	....	5† 5	5 38	5 48	5 53	6 48	6 54
2	9	Shipton-on-Cherwell Halt .. ,,	69 R	7 29	..	8 48	..	11 21	—	..	12 44	1 16	3 6	..	4 16	..	—	5 44	5 54	5 59	6 54	7 0
3	57	Blenheim & Woodstock .. .... arr.	129 R	7 33	....	8 53	....	11 25	12 10	....	12 48	1 20	3 10	....	4 20	....	5†13	5 48	5 58	6 3	6 58	7 4

Up Trains.

STATIONS.	Ruling Gradient 1 in	B Auto Mixed.		B Auto.		G Engine and Van. RR SX	B Auto.		B Oxford Auto.	B Auto. K SO	B Oxford Auto. K SO	B Auto.		K Freight K		B Auto.		B Auto.		B Auto.	
		a.m.		a.m.		a.m	p.m.		p.m.	p.m.	p.m.	p.m.		p.m.		p.m.		p.m.		p.m.	
Blenheim & Woodstock .. .... dep.	—	7 58	....	9 30	....	11†30	12 22	....	12 58	12 58	1 25	3 48	....	4 40	....	5 22	....	6 20	....	7 10	....
Shipton-on-Cherwell Halt .. ,,	129 F	8 4	..	9 35	..	—	12 27	..	1 3	1 3	1 30	3 53	..	—	...	5 27	..	6 25	..	7 15	..
Kidlington .... arr.	69 F	8 11	....	9 40	....	11†38	12 32	....	1 8	1 8	1 35	3 58	....	4 50	....	5 32	....	6 30	....	7¶20	....

K—Suspended. N—On Saturdays commencing 23rd May, 1953, will start at 6.54 p.m. V—Commences 23rd May, 1953. ‡—May run as a Mixed trip when necessary at point-to-point times shewn for other Mixed trips. ¶—Thence empty to Oxford at 7†25 p.m.

Figure 25: The last Woodstock branch working timetable, for 1953, just before closure.
Courtesy Laurence Waters; Great Western Trust collection Didcot.

Plate 51: The last photograph of the halt, which had now closed to passenger traffic, taken in 1957 just before the track was lifted. Just one image from the Industrial Railway Society's vast photographic collection showing on the extreme left, the houses at Bunkers Hill and their proximity to the halt.
Courtesy Industrial Railway Society; photograph by R.M. Casserley.

As stated previously, the installation of the exchange sidings for the Shipton cement by the GWR had already begun in late 1927, before any construction of the actual works had started. They consisted of three long sidings, as can be seen in the track plan in *Figure 18* and the frontispiece photograph. The GWR Company's boundary wire fence which included the exchange sidings, can be seen in *Plate 103*. It was also recorded that there was a space allowed for a fourth line to be laid later if it was deemed necessary.

The all-important new signal box and track connection into the Oxford to Banbury main line was next to be constructed. The name of this new signal box was Bletchington Cement Sidings and the superb aerial photograph in *Plate 55*, shows the position of this box as a small white roofed building in the lower right-hand corner of the image.

COPY.

S.R.11479.

MINISTRY OF TRANSPORT,
7, Whitehall Gardens,
London, S.W.1.

24th January, 1929.

Sir,

I have the honour to report for the information of the Minister of Transport in accordance with the Minute of the 9th December 1927, that I made an inspection on the 15th instant of the new works at Bletchington on the Oxford - Banbury double line north of Kidlington on the Great Western Railway.

New sidings for the Oxford and Shipton Cement Company have been laid on the west side of the railway to which access at the south end is provided over a through crossing from new trailing points on the up line. Single slips on this crossing provide a crossover between the up and down line. At the north end access to the sidings is from a new set of trailing points on the down line.

The main line connections north and south are separated by a distance of 636 yards. The necessary distant, home and starting signals for each road for block control of traffic have been provided. A new signal box (Cement Sidings) on the east side of the railway, situated intermediate between the loop terminals, forms a new block post on the line, and contains 18 working levers, of which two are detonator placers, and 6 spaces. There is an old track circuit at the north end in advance of the down starting signal. At the south end there are two new track circuits, one on the up line between the up home and up starting signal, and the other in rear of the down home signal. The interlocking in the frame is correct, and the necessary arrangements, including repetition signals etc. are satisfactory.

The signalman drew attention to the position in which the detonator was placed on the down main line. In accordance with the usual practice the position is in front of the box. The objection to placing it in front of the box in this instance was because it is in rear of the up home signal and may sometimes be used by signalmen when it is unnecessary. I do not, however, see that there is any real necessity for altering the arrangement and therefore do not make a requirement.

I recommend that final approval be given to these new works.

I have the honour to be,
Sir,
Your obedient Servant,

(Sgd) J.W. Prigle.
Colonel.

Plate 52: This is an interesting letter from Colonel Pringle, who was the Ministry of Transport inspector, regarding the track-work and signalling for the new GWR Bletchington cement sidings signal box at Shipton cement works, dated 24th of January 1929. This was a long time after the work was carried out on the 9th of December 1927, as recorded in the GWR New Works registers 1891-1951.
Courtesy National Archives; File MT29/85/41.

The signal box was opened on the 7th of November 1927 and closed on the 16th of September 1968.

The construction was a typical GWR design with brick bottom half, wooden top half and was 21 ft 2 ins long, 11 ft 2 ins wide and 8 ft high, containing 24 levers of which eighteen were working (that included two which were used for detonators) and six spares. The lever operation is shown in the signalling diagram in *Figure 26.*

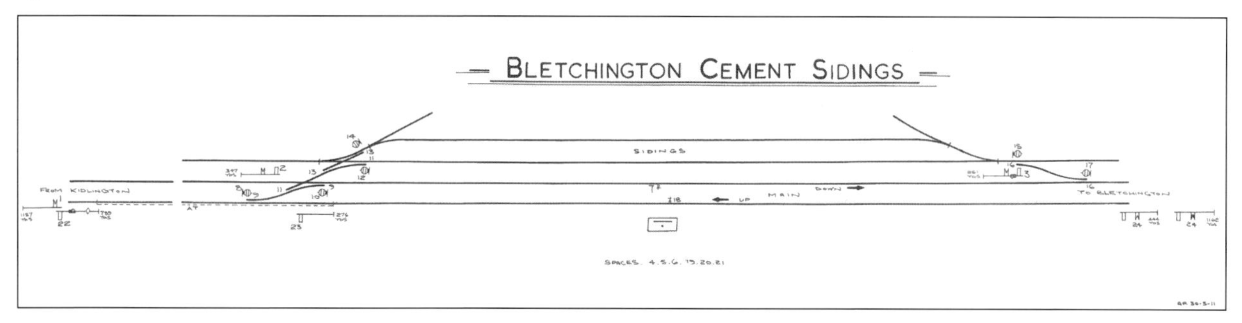

Figure 26: The signalling diagram for the new Bletchington cement sidings signal box. The opening times were 11am until 5.10pm weekdays. *Courtesy Chris Turner; Signalling Record Society.*

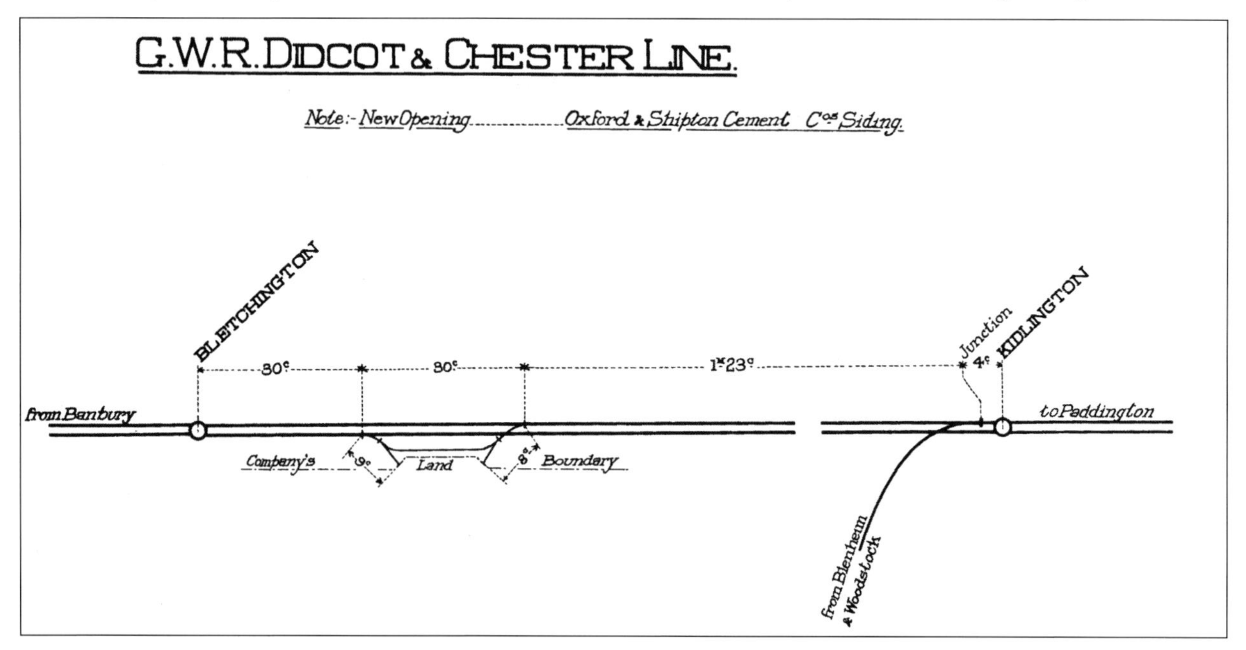

Figure 27: This is a diagram produced by the Chief Civil Engineer's office for new facilities, to provide accurate distances for the Goods Department for freight charging purposes.
Courtesy Chris Turner; Signalling Record Society.

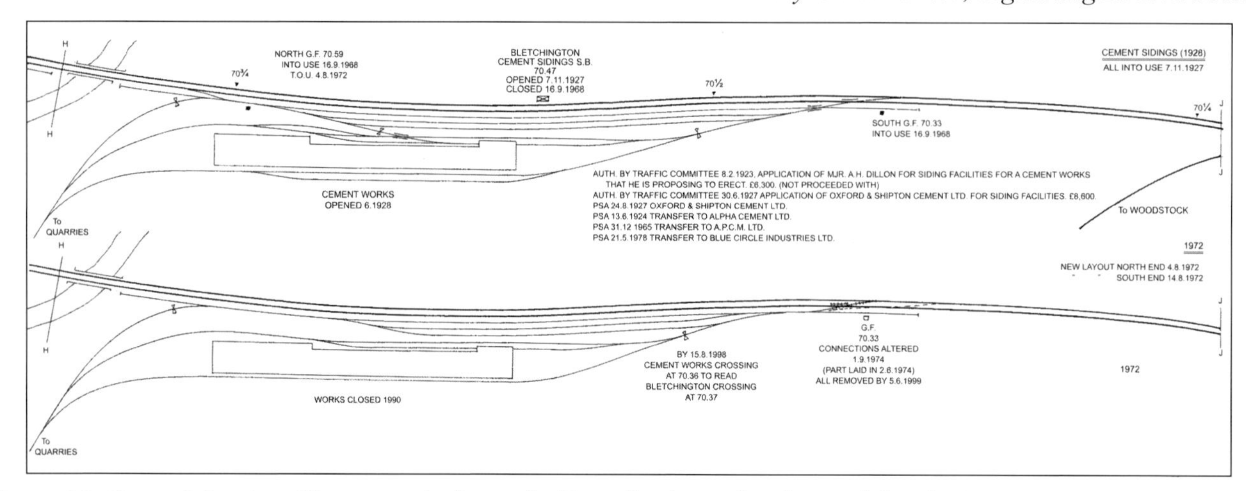

Figure 28: Part of the signalling records drawn by Tony Cooke and re-lettered for clarity.

Private and not for Publication. **Notice No. 384.**

GREAT WESTERN RAILWAY

(For the Company's Servants only).

SIGNAL ALTERATIONS—BLETCHINGTON

Opening of New Signal Box, Cement Sidings Signal Box, Bletchington

(Situate between Kidlington and Bletchington).

On MONDAY, NOVEMBER 7th, between the hours of 8.0 a.m. and 4.0 p.m., or until the work is completed, the Signal Engineer will be engaged in bringing into use a new Signal Box fixed on the Up Side of the Up Main Line, half a mile south of Bletchington Station, which will be known as Cement Sidings Signal Box, Bletchington, to control trailing connections for the Cement Company's new Sidings to the Up and Down Main Lines, and the following new Signals will be introduced :—

	Form	Description	Position	Distance from Box
A		Down Main Distant.	Up Side of Up Main Line.	1,157 yards.
B	(1) (2)	1. Down Main Home. 2. Down Main Distant for Bletchington. Train indicator at this Signal.	Down Side of Down Main Line.	347 yards.
C	(1) (2)	1. Down Main Starting. 2. Down Main Inner Distant for Bletchington.	Down Side of Down Main Line. (Between Down Main and Sidings).	261 yards.
D	(1) (2)	1. Up Main Inner Home for Bletchington. 2. Up Main Distant for New Box.	Down Side of Down Siding (By Bletchington Box).	851 yards.
E	(1) (2)	1. Up Main Starting for Bletchington. (Already in use). 2. Up Main Inner Distant.	Up Side of Up Main Line.	333 yards.
F		Up Main Home.	Up Side of Up Main Line.	276 yards.
G		Up Main Starting. (Train Indicator at this Signal).	Up Side of Up Main Line.	789 yards.

Discs will be fixed at the points leading from Sidings and at the points of Crossover Road.

The existing Down Main Distant and Up Main Inner Home for Bletchington will be taken out of use.

During the time the work is being carried out, Distant Signals Nos. 1 and 29 in Bletchington Box will be disconnected and placed at danger.

The Block Telegraph arrangements will be as follows :—

Double Line Disc Block, Bell and Box to Box Telephone communication will be provided with Bletchington Signal Box and Kidlington Signal Box.

A Block Switch will be provided. A telephone will be provided on the Oxford—Heyford Circuit.

TRACK CIRCUITS.

Signal Box.	Length and Position of Track.	Position of "Train Waiting" Indicator.	Signals Electrically Locked when Track is occupied.	Whether Block controlled.	Whether "Vehicle on Line" Switch provided.
Cement Sidings.	513 yards to rear of Up Main Starting Signal.	At Up Main Starting Signal.	Up Main Home.	No.	No.
Do.	500 yards to rear of Down Main Home Signal.	At Down Main Home Signal.	Kidlington Down Main Advanced Starting Signal.	Yes.	No.
*Bletchington.	550 yards to rear of Down Main Home Signal.	At Down Main Home Signal.	Cement Sidings Down Main Starting Signal and Kidlington Down Main Advanced Starting Signal when Cement Sidings Box is switched out.	Yes (only when Cement Sidings Box is switched out).	No.

*Replaces existing Track Circuit which electrically locks Kiddlington Down Main Advanced **Starting Signal only.**

The new Signal Box will be open on week days only between the hours of 9.30 a.m. and 5.30 p.m. until further notice.

The work will be carried out by **Inspectors Virgo and Lascelles,** of Reading.

District Inspector Allsop to arrange Flagmen, and see that all necessary precautions are taken to ensure safety in accordance with Rule 71 of the General Rules and Regulations.

The receipt of this Notice to be acknowledged by first train.

Birmingham,
October 29th, 1927.

A. BROOK,
Superintendent of the Birmingham Division.

Chance & Bland Ltd., Printers, Gloucester.

Figure 29: This is GWR notice 384, covering the opening of the new Bletchington cement sidings signal box. *Courtesy Tony Cooke collection.*

Private and not for Publication. **Notice No. 29**

GREAT WESTERN RAILWAY

(For the use of the Company's Servants only).

BLETCHINGTON.

On **WEDNESDAY, FEBRUARY 15th,** the Signal Engineer will be engaged from 10.0 a.m. to 3.30 p.m., bringing into use the following new Signal :—

Form	Description	Position	Distance from Box
1, 2	1. Up Main Home for Bletchington Station (already in use). 2. Up Main Distant for Cement Sidings.	Up side of Line.	1,152 yards.

The existing Up Main Distant for Cement Sidings will become the Up Main Intermediate Distant.

During the time the work is being carried out, Distant Signals affected, No. 24 and No. 1 in Bletchington Box will be disconnected and placed at "Danger."

Inspector Allsopp to make all arrangements for the safe working of the Line (including the appointment of any Groundmen) in accordance with Rule 71.

A. BROOK,
Superintendent of the Birmingham Division.

Birmingham, February 4th, 1928.

The receipt of this Notice to be acknowledged by first Train.

..............Station..............1928

Received copy of **Mr. Brook's Notice No. 29,** re Bletchington.

...........................(S).

B.H. 2990 Chance & Bland Ltd., Gloucester.

Figure 31: This notice No. 29 is for additional work to be carried out to the Bletchington signalling in February 1928. Courtesy Tony Cooke collection.

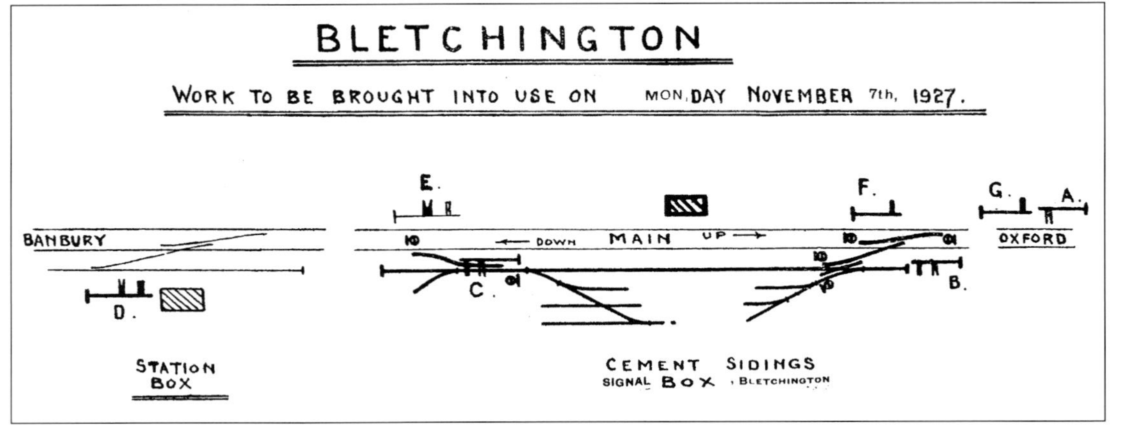

Figure 30: This is a diagram of the new track and signalling to be brought into use on a specific date, which supports the detail in the signalling notice No. 384, issued to the staff. It is required for operational purposes especially for the train crews. Dated October 29th of 1927 for the work to be carried out by November 7th 1927. Courtesy Tony Cooke collection.

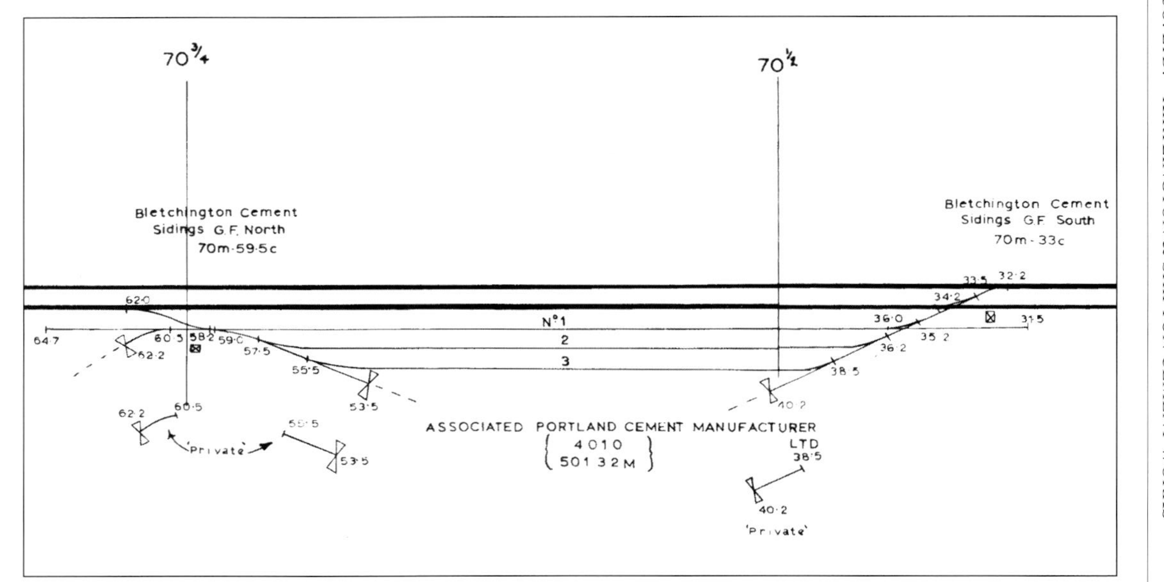

Figure 32: After the Bletchington cement sidings signal box was closed, the track work leading into the Shipton cement works exchange sidings was controlled by two ground frames – one at each end of the exchange sidings and the above figure shows the ground frame positions relative to the main line mile posts. Courtesy Chris Turner collection.

BLETCHINGTON.

ALPHA CEMENT CO.'S SIDING.

1. These Sidings are situated on the Down side of the Line between Kidlington and Bletchington, and have connections with the Up and Down Main Lines, controlled from the Cement Sidings Signal Box.

2. The Oxford and Shipton Cement Co. perform their shunting work at the Sidings with a private locomotive, accompanied by a Shunter employed by the firm.

3. This locomotive works into and out of the three Sidings (Nos. 1, 2 and 3) parallel with and adjoining the Down Main Line, No. 1 Siding being the road next to the Down Main Line.

4. No. 1 Siding must, as far as possible, be kept clear for the reception of a train calling at the Sidings to attach or detach traffic.

5. The Firm's private locomotive will cease operations in Nos. 1, 2 or 3 Sidings when requested by the Great Western Signalman, Guard or Checker to do so for a Great Western train to work.

6. Guards of trains calling to pick up or put off traffic, or to perform shunting operations at the Sidings must satisfy themselves that all is clear before commencing operations, and have a clear understanding with the man in charge of the Firm's locomotive if it is about.

7. The Cement Co.'s locomotive must not under any circumstances be allowed to go on to the Up or Down Main Lines.

Figure 33: Extract from the 1938 Great Western Railway Working Appendix relating to the operation of the Shipton (Bletchington) cement works. Courtesy Laurence Waters; Great Western Trust collection Didcot.

A summary of the working timetables is included and, in the Autumn of 1929, the Company sidings were served by the 3.30am goods, an empty goods wagon train from Old Oak Common to Leamington Spa, which ran from Tuesday to Saturday. The train served the company sidings from 7.55am to 8.15am, presumably leaving empty wagons from the 'wagon pool' at Old Oak for loading.

Later in the day another Leamington bound goods train (the 7.55pm from Oxford) called at the Company sidings between 8.12pm and 8.32pm to collect the loaded wagons.

During weekdays, in the up direction, the 1.15 pm Heyford to Oxford goods train called in at Shipton works from 1.53pm to 2.13pm. Then the 2.25pm Leamington to Oxford coal and goods train called in from 5.53pm to 6.08pm leaving the vital coal supplies for the works.

Plate 53: Only one picture has been discovered of a steam hauled train leaving Shipton cement works (image looking south), with a local pick-up class 7 express freight goods with 56xx class 0-6-2T No.6697. The train has just picked up a sheeted hopper wagon, possibly full of cement (the first wagon), with the rest empty coal wagons and now moving out from the Shipton-on-Cherwell cement works onto the mainline and onwards towards Birmingham and beyond. This locomotive was built in 1928 by Armstrong Whitworth in Newcastle and was based at several depots, but by 1964 was at Leamington Spa and in June 1965 arrived at Banbury shed. Now in preservation, it is on static display at GWS Didcot Railway centre.

Courtesy of Steve Banks collection, photographer R. Brough, R.K. Blencowe Archive

The afore-mentioned services show a pattern of traffic distribution north via Leamington Spa and south via Oxford. Almost certainly because of the development of the marshalling yard in the 1930s at Banbury, the sorting and onward transit of the north bound traffic became centred there instead of Leamington.

As an example of the north bound changes, in the Autumn of 1946, a 9.00am local goods train from Banbury was scheduled to terminate at the Shipton sidings at 12.08pm returning to Banbury at 12.58pm. Then later the 1.45pm Banbury to Oxford called at Shipton from 3.53pm to 4.08pm and whilst this train served the company works, it was also refuged there to allow other trains to pass.

The 1960 working timetable is included in full in *Figure 34 and 35*.

A50 WEEKDAYS BANBURY, OXFORD, NEWBURY AND READING

UP		K	K	H	K	K	K	F	H	K	H	H	K	K	H	H
			To Abingdon	8.50 am SX Q Eastleigh to Didcot	To Morris Cowley	To Slough		5.55 am Stoke Gifford to Woodford L.M.R.	To Tonbridge S.R.			To Westbury		To Morris Cowley	12.30 pm Basingstoke to Sonning	7.5 am Stoke Gifford to O.O.C.
		9A16	9A00	8V26	9A41	9A69	9A66	7M09	8O06	9A66	8A50	8B57	9A66	9A41	8V23	8A53
		SX	SX	SX Q	SX		SX	SX	SX	SX	SX	FSX	SO		Q	MSX
		am	am	am	am	am	am	am	PM	PM	am	am	am	PM	PM	PM
BANBURY GEN.... dep	1	..	..	..	..	..	..	..	..	..	9 55	10 30	..	..	..	..
Aynho Junction...............	2	......	......	......	......	......	......	......	......	......	10 10	10 49	......	......	......	......
Cement Sidings...	3	..	..	..	..	..	..	..	..	..	12N10	11 13	..	..	..	..
Kidlington	4	......	Sandford arr. 11R0 am	......	......	......	......	......	......	......	1N 0	......	......	......	......	......
Yarnton	5	..		..	..	..	..	..	..	..	..	..	..	..	..	..
Wolvercot Junction.........	6	......		......	......	......	......	......	......	......	......	......	......	......	......	......
OXFORD ... ... arr	7	..		..	..	..	..	..	..	..	..	11*31	..	..	..	..

BANBURY, OXFORD, NEWBURY AND READING WEEKDAYS A51

H	K	C	K		H	H	H	F		H	G	K	D	H	—	H	K	H
6.10 am Wellingboro'	10.6 am Uffington	6.25 am Tavistock Jn. to O.O. Common	To Didcot		3.15 am Alex. Dock Jn. to O.O.C.	6.5 am Severn Tun. Jn. to O.O.C.	11.50 am SX, 12.35 am MO Neath	11.37 am WFO Swindon (Rodbourne Lane) to Yarnton		6.55 am Severn Tunnel Jn.	1‖15 pm Oxford Shed to Morris Cowley L E		9.30 am Oxley Sidings to Moreton Cutting		12.45 pm Swanbourne L.M.R.	To Eastleigh	1.30 pm Welford Park	7.35 am Severn Tun. Jn. to Hanwell Bridge Sidings
8V64	9A16	4A60	9A16		8A48	8A59	8A55	7A30		8A61	0A41	9A28	5A57	8V24	8V68	8O20	9A42	8A63
SO	SX	SX	SX		MO	SX		WFO		MSX	SX	SO	SX	SX		SX Q	MW FO	MSX
PM	PM	PM	PM		PM	PM	PM	PM		PM	PM	PM	PM	PM	PM	PM	PM	PM
..	..	..	..	..	..	..	..	..	..	..	..	..	..	..	..	..	..	..
......	......	......	......	......	......	......	......	......	......	......	......	......	......	......	......	......	......	......
..	..	..	..	..	..	..	..	..	..	..	..	1 10	..	..	..	..	..	..
......	......	......	......	......	......	......	......	......	......	......	......	......	......	......	......	......	......	......
..	..	..	..	..	..	..	11 46	..	..	..	..	..	1 48	..	..	..	..	..
......	......	......	......	......	......	......	......	......	......	......	......	......	GL	......	......	......	......	......
..	..	..	..	..	..	..	..	..	..	..	..	..	..	..	..	..	..	..

Figure 34: The 1960 working timetable extract for the cement works sidings on the main line from Banbury to Oxford. *Courtesy Chris Potts collection.*

A34 WEEKDAYS READING, NEWBURY, OXFORD AND BANBURY

DOWN		F	K	K	H	G	D	D	F	G	F	—	K	F	H	H
		Empties to Banbury Jn.	7.0 am Slough	8.25 am Didcot	9.50 am Didcot to Eastleigh	E B V	4.15 am Taunton to Banbury Jn. MSX, Bordesley Jn. MO	4.15 am Taunton to Banbury Jn.	3.50 am MSX, 4.10 am MO, Stoke Gifford to Banbury Jn.	L E	Empties to Banbury	8.44 am Feltham S.R.	8.50 am Henley-on-Thames	5.55 am Stoke Gifford to Woodford	1.30 am Snodland	8.40 am Woodford to Stoke Gifford
		7H24	9A66	9A29	8O19	0A11	5H35	5H35	7H28	0A16	7H26	8V07	9A66	7M09	8V06	8V77
OXFORD ... ... arr	47															
... ... dep	48	10 13	10 39	10 20	..	10 25	10 49	10 45	11 5	..	11 35	..	..	12 20	..	..
Wolvercot Junction	49							11* 5	11*57							
Yarnton	50	..	10 50	10 35	..	..	..	..	..	..	..	..	..	..	..	..
Kidlington	51															
Cement Sidings...	52	10 32	..	..	..	10†50	11N 0	11 14	12N25	..	11 50	..	..	12 40	..	..
Aynho Junction..............	53	10 55					11 27	11 50	12 51		12 42			1 3		
BANBURY GEN. arr	54	11 6	..	..	..	..	11 48	12 15	1 9	..	12 56	..	..	1T30	..	..

READING, NEWBURY, OXFORD AND BANBURY WEEKDAYS A35

	K	H	K	K	H	K	H	C	F	G	F	H	F	F	F	—	F	—	H
	To Oxford Rewley Rd.	8.50 am Eastleigh	10.30 am Thatcham	11.50 am Morris Cowley	10.6 am Uffington	To Welford Park	7.55 am Banbury to Westbury	3.0 am TO Croes Newydd (South Fork) to Swindon	10.55 am Acton to Stoke Gifford	E B V	To Severn Tunnel Jn.		10.50 am Old Oak Com. M.P.D. to Rogerstone Empties	11 15 am Acton to Severn Tunnel Jn.	11.37 am Swindon (Rodbourne Lane) Yarnton WFO	5.50 am Tonbridge	12.5 pm Acton to Severn Tunnel Jn.	10.4 pm FSX Angerstein Wharf	10 30 am Banbury to Westbury
	9A50	8V26	9A16	9A28	8A51	9A84	8B56	4B38	7B33	0A39	7T49	8H26	7T51	7T55	7A30	8V17	7T55	8V11	8B57
47												2 43							
48	11 50	..	..	..	..	1 15	..	..	..	..	..	..	..	..	2 0	..	..	..	..
49																			
50	Oxford North Jn. pass 11.52 am	..	..	..	..	Oxford North Jn. 1.17 pm	..	..	..	..	..	..	..	..	2 10	..	..	..	..
51												3H19							
52		..	..	..	..		..	..	..	..	..	4 15	..	..	..	..	..	..	..
53												5 52							
54		..	..	..	..		..	..	..	..	..	6 16	..	..	..	..	..	..	..

Figure 35: The 1960 working timetable for the other direction of the Oxford to Banbury mainline.
Courtesy Chris Potts collection.

Plate 54: An immaculate Warship class No. D851 TEMERAIRE with the 6M07 freight train waiting to leave Shipton-on-Cherwell cement works sidings with what looks like an empties coal train after the first wagon, which looks like a bogie bolster wagon. *Courtesy Laurence Waters, Great Western Trust collection, Didcot.*

Chapter Six

Change of ownership – ALPHA CEMENT LTD, APCM LTD, BLUE CIRCLE INDUSTRIES LTD, and finally BLUE CIRCLE INDUSTRIES PLC.

The sequence of the legal ownership of the new Shipton-on-Cherwell cement works was as follows:

* Oxford & Shipton Cement Ltd., from its construction and opening until July 1934.
* Alpha Cement Ltd., from August 1934 until 31st of December 1965 (but controlled by APCM Ltd., from March 1938).
* APCM Ltd., from 1st of January 1966 until 31st May 1978.
* Blue Circle Industries Ltd., from 1st of June 1978 until 7th of October 1981.
* Blue Circle Industries PLC, from 8th of October 1981 until closure in 1987 (this was because of a change in legislation, requiring public limited companies (PLC) to be differentiated from private limited companies).

The Times newspaper of the 29th of March 1934 under the heading – **Oxford & Shipton Cement – Offer of Shares:**

> An agreement has been entered into whereby British and Allied Investments Corporation undertakes to purchase all the issued shares (450,000 of £1 each) of the Oxford & Shipton Cement Ltd., at a price of 28s a share plus an amount equal to 4% per annum, less tax, calculated on the nominal amount of the shares from January 1st, 1934, down to date of payment. The contract is provisional only to the condition that the holders of not less than 70% of the issued (or such smaller percentage as the purchasers may be willing to accept) agree to sell their shares on these terms. If sufficient acceptances are received it is intended that the company shall acquire the whole of the share capital of the Alpha Cement Company – namely 280,000 Ordinary shares of 10s each and 110,000 seven percent preferences shares of £1 each. The Ordinary shares to be acquired by exchanging one fully paid £1 share for two Ordinary 10s shares of the Alpha Cement Company, and Preference shares at the price at which they are redeemable – 22s 6d a share, plus an additional sum equal to 7%, per annum from 1st of August, 1933, to date of completion, payable in cash.
>
> To carry this into effect it is proposed to increase the capital of the Oxford & Shipton Cement Company by 200,000 shares of £1 each and to create £300,000 four and half per cent Debenture stock, in respect of which the company will receive a net 97½%, less expenses. The authorised capital of the company will then be 700,000 shares of £1 each, of which 617,500 shares will be issued and fully paid up and as it is intended that the bank loans of the Alpha Cement Company, which stood at £95,000 on December 1st last, and of the Oxford & Shipton Cement Ltd., which now stands at £42,000, shall be paid off, the only prior charge will be £300.000 – four and a half per cent Debenture stock.
>
> It is stated that the Alpha Cement Company, which only recently commenced partial production, will be ready for production at its full capacity of approximately 100,000 tons per annum in April this year.
>
> Four of the five directors of the Oxford & Shipton Cement Ltd. will retire from the Board and to compensate them the loss of office, 27,500 shares are to be issued to them for cash at par. These shares will be acquired by British and Allied Investments Corporation at 28s a share and the difference of £11,000 will be divided between the directors.

Then, by the 20th of April 1934, an advert appears for the Oxford & Shipton Cement Limited announcing the terms of a Debenture Stock for £300,000 *(see below).*

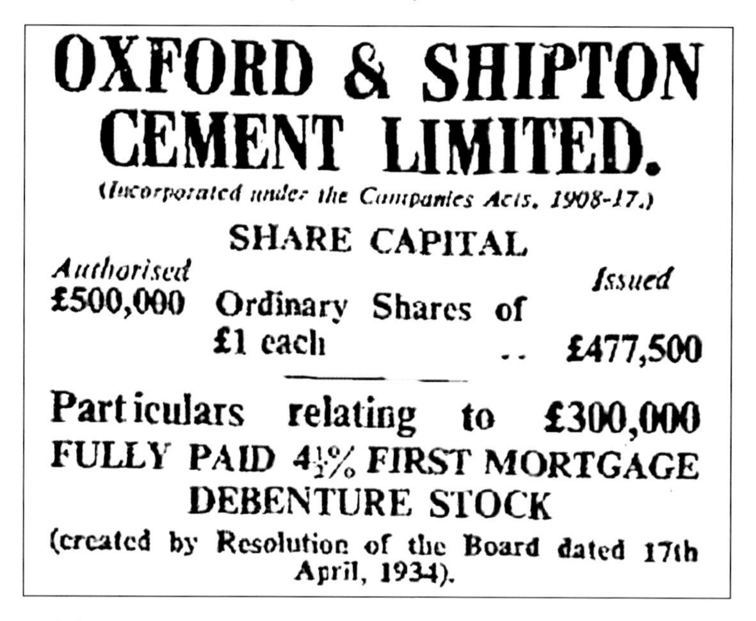

OXFORD & SHIPTON CEMENT LIMITED.

(Incorporated under the Companies Acts, 1908-17.)

SHARE CAPITAL

Authorised		Issued
£500,000	Ordinary Shares of £1 each ..	£477,500

Particulars relating to £300,000 FULLY PAID 4½% FIRST MORTGAGE DEBENTURE STOCK

(created by Resolution of the Board dated 17th April, 1934).

Figure 36: Facsimile copy of the Times newspaper 1934 advert. *Courtesy Chris Down collection.*

It seems a sad coincidence that Lord Dillon (1875 to 1934) had sadly passed away (perhaps due to all the stress associated with the sale and setting up of this new company) so very soon after all this activity and changes in the company affairs and his obituary was placed in The Times newspaper on 28th of May 1934.

Lord Dillon had died about ten days before this and was due to take up a new position as Chairman of an investment company in London. He had never had ideas of grandeur and for family reasons went on to the Stock Exchange. However, besides stockholding, he went into the cement industry and in later years built the great and successful cement company at Shipton-on-Cherwell. He had just very recently weathered a sale of the very successful Oxford & Shipton Cement Ltd. and was looking forward to the new opening in his life.

Alpha Cement Limited was founded in 1933 by Albert Younglove Gowen (1883 to 1964), who, like his father, had been associated with the American Lehigh Portland Cement Corporation, which was one of the largest American cement manufacturing companies. Gowen apparently came to England on a holiday and found that there could be great opportunities in reorganising the English cement industry and so decided to stay.

Gowen quickly set up his own company – ALPHA CEMENT LTD. – which was a name already established in America and used this to buy, in September 1933, the Rodmell cement works, Sussex. By March 1934 he was well on the way to owning Oxford & Shipton Cement works. Then he followed that in July 1934 with the purchase of Cliffe (or Thames) cement works, Kent. A year later, in 1935, the Kirton Lindsey works, Lincolnshire and finally the Metropolitan cement works, Essex in May 1936, which was still under construction at this time – making five acquisitions of British cement plants in all.

It is interesting to note that Albert Younglove Gowen was appointed Managing Director in 1934 of the Oxford & Shipton Cement Ltd., before the change to Alpha Cement Ltd. However, when Alpha Cement Ltd. was eventually acquired by APCM, he became a director of APCM and that continued until 1947.

Plate 55: This aerial photograph taken in the 1950s shows the Shipton cement plant as depicted in the diagrammatic plan seen in Figure 19, with the new tall white chimney stack dominating the area. The new GWR Bletchington cement sidings signal box can be seen on the right of the photograph alongside the Oxford to Birmingham main line (small white roofed building to the bottom right of the photograph).

Author's collection, courtesy of the late J.H. Russell collection.

Gowen was a great fan of calcinators and equipped four out of the five works with them – but not Kirton Lindsey.

The size of this company was now about the same as Tunnel Portland Cement and second in size to APCM. However, in March 1938, having struggled as an independent organisation, he sold the company to Tunnel Cement (26%) and APCM (74%) with himself a director of both companies.

The Alpha Cement Company Ltd. title was retained until 1949 when APCM acquired 100% of the share ownership and integrated Alpha's business into its existing structure. Even then the Alpha company name remained until 1964, when it ceased to trade, but continued to exist on the companies register until at least 1981.

As far as Oxford & Shipton works was concerned, it was reported in 1938 that one of the original rotary kilns (No.2) was removed and replaced by one of the same length but an increased diameter. At the same time a third kiln of 9ft 8ins in diameter was also added. APCM and associated companies became better known by their trade mark (a blue circle) to the point, that in 1978, the trade mark became the group's legal identity – Blue Circle Industries. The Group by then had become an international group of subsidiary and associated companies, primarily engaged in the manufacture and distribution of cement and allied products, decorative finishes, the extraction of mineral and the merchanting of building materials. The Group also manufactured and marketed ceramics and acrylic sanitary-wares and bathroom products and boilers through various subsidiaries.

Plate 56: Included in these images are the logos that the Blue Circle company used on their various products and on their road transport fleet.

Authors collection all obtained from an international auction site and reproduced under the Creative Common licence.

The origin of APCM goes back to 1900, when 30 single cement works companies were merged into the Associated Portland Cement Manufacturers (1900) Ltd.

It was the largest cement manufacturer in the United Kingdom and one of the largest cement manufacturers in the world, with operations in all five continents, including interests in 45 cement works and sales of 21 million tons of cement in 1981 alone.

Plate 57: A fine view, looking roughly southwards from beside the ex-GWR main line. On the right of the photograph can be seen the conveyor carrying the crushed stone up from the quarry to the raw mills and then the three so-called 'doctor tanks' which contained various minor additives required to fine-tune the chemistry of the raw materials. Next left is the single tall main chimney (built 1948 and locally called 'Smokey Joes') with the kiln house behind and the electrostatic precipitators intended to reduce dust emissions; the latter were only installed c1962. The ex-GWR main line can be seen on the extreme left.

Courtesy Oxfordshire County Council, Oxford History Centre ref; POX015870.

Plate 58: Oxfordshire and, particularly, the area around the City of Oxford was not usually associated with any heavy industry, but at Shipton-on-Cherwell (only six miles approximately from the city centre) any visitors arriving by train from the Midlands would have been greeted by the above sight – a huge white chimney, large buildings, and an area of wilderness. Note also the locomotive and three wagons near the Mitchell wagon lift right in the middle of the photograph. *Courtesy of George Reeves, Irwell Press; photograph by John Bonser.*

Plate 59: Locomotive No. 2 ALBION, a 0-4-0ST by W.G. Bagnall Ltd. (WB 2178/1921), which came to Shipton from Rodwell cement works, Sussex, seen here outside the only engine shed on the Shipton-on-Cherwell cement works internal railway system. Another one of the locomotive steam fleets is seen inside the shed.
This early view of the shed shows it was constructed with a small 3 feet high brick base at ground level, with corrugated iron sides and asbestos roof and could hold two locomotives inside. It had an overhead crane used for wheel changes and workshop machinery along the wall inside. The main door appears in this photograph to be a canvas sheet. It was manned by two qualified fitters, who acted as drivers as well.
Courtesy The Industrial Railway Society.

Plate 60: Another later view of the locomotive shed in June 1967 with WB 2178/1921 No. 2 peeping out from a very dark interior. It looks as though the shed has received new cladding on each side of the opening but still no door, only curtaining for the doorway. Also, a proper coal bunker has been added on the right hand side and the track on the left of the shed seems to have been lifted. *Courtesy John Scholes collection.*

Plate 61: A line up of new Alpha Cement Ltd. delivery lorries.
Courtesy Oxfordshire County Council, Oxford History Centre; ref POX0130647.

Plate 62: A low quality photograph showing the entry into the Shipton-on-Cherwell works, reproduced from a booklet produced by Alpha Cement to celebrate 50 years of operation.
Courtesy collection of John Oxford.

The Oxford Shipton Works as part of the international Blue Circle Group, was but one element in an overall corporate strategy that, in the UK, became centred around a major change in the manufacturing process. Cement manufacture in the nineteenth and early twentieth centuries used what was known as the "wet process" *see Figure 20*. Typically, raw materials were mixed in the proportions 3 to 4 parts limestone or chalk to 1 part clay or shale. The manufacturing process had evolved to handle the available raw materials in the South East, chalk and clay, each having a high (15-25%) moisture content. Directly mixing damp rocks in the right proportions was not technically possible for decades; the only way to mix them was to add water to each, turn them into slurries and then blend the chalk and clay slurries in the correct proportions. The correct mixture was therefore also in slurry form and, when introduced into the kiln, a large amount of fuel was needed to first burn off the added water, before it began to heat the mixture (up to about 1450°C) to convert it into cement.

During the Second World War, the production capacity of Shipton was rated at 5,150 tons per week – over a 50-week year, 257,500 tons (equal to about 270,000 tons of cement). During the 1942-45 period, actual production fluctuated widely between 30% and 80% of that nominal figure.

The next output snapshot comes in the 1964 when the factory manufactured and despatched both bagged and bulk cement, plus small amounts of cement clinker and raw limestone for use elsewhere. The means of transport, however, was strongly biased in favour of road. It is reported that some 500,000 tons of cement were despatched by road and only 22,000 tons by rail (both bulk and bagged). Other minor despatches were solely by road and consisted of 33,000 tons of clinker and 16,500 tons of limestone. Most incoming traffic to the works was rail-borne and consisted of 69,000 tons of coal and 17,500 tons of gypsum. Very small amounts of both cement and clinker came in by road and certainly by 1977, all cement despatches were by road.

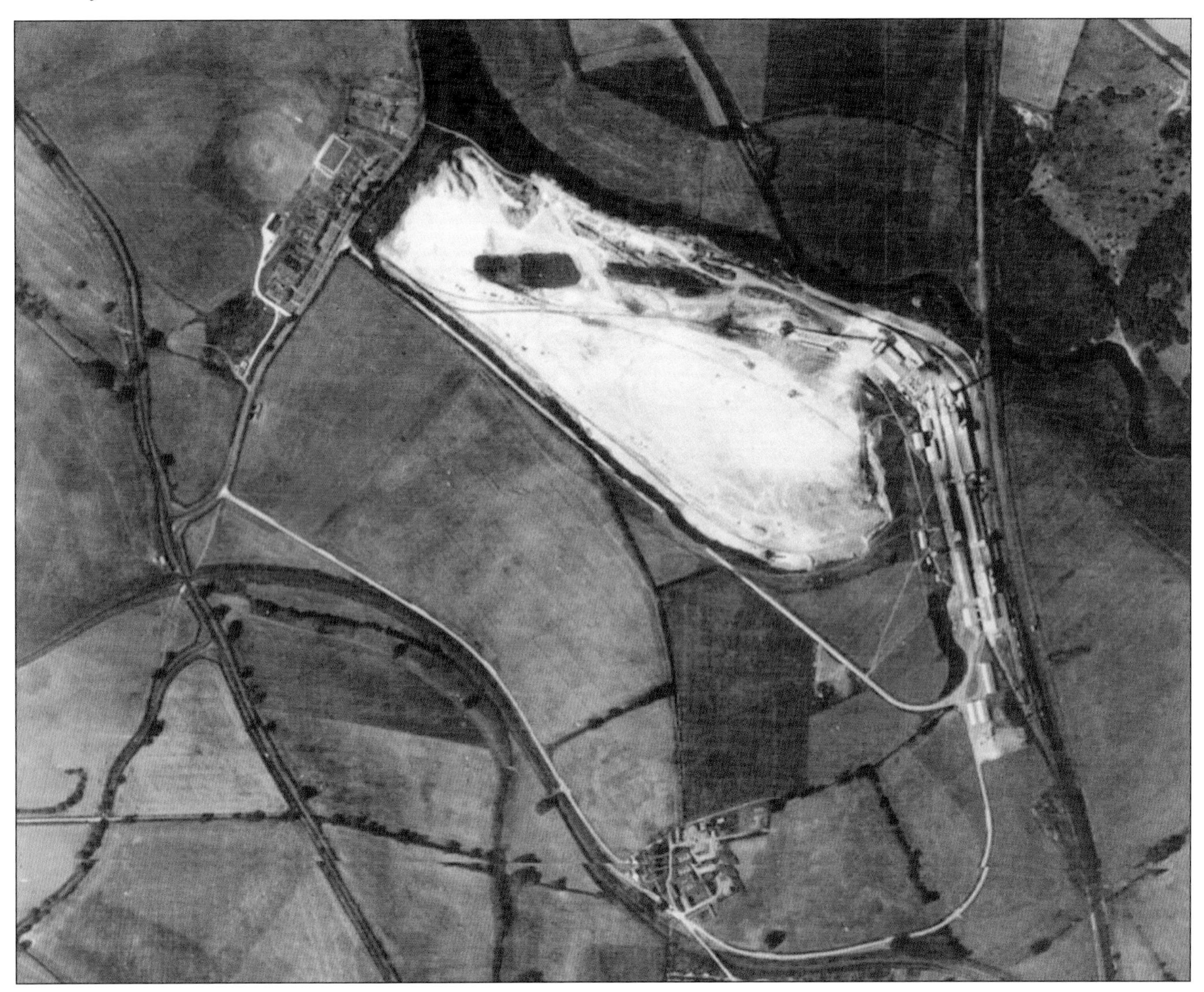

Plate 63: A great 1949 aerial view of the Shipton cement works and quarry. Note how the new in-road follows the Woodstock branch line near the bottom of the photograph. The BR Oxford to Banbury main line running up the right hand side of the image. The Bunkers Hill housing can be seen on the top left hand corner and the internal rail tracks can be seen on the quarry floor. The original works access road, from in front of the Bunkers Hill houses, can still be seen, but the new replacement access road, can also be seen curving around what became the ultimate extent of the quarry face.

Reproduced courtesy National Library of Scotland under the Creative Commons Attribution licence.

Although Shipton's raw materials, limestone and shale, were low-moisture rocks, there was no way of blending them without adding water, so that Shipton used a wet process as well. By the 1950s, it had become possible to blend such rocks without adding water – called the "dry process" – and Blue Circle embarked upon building a series of new regional dry process factories across the UK. By the 1960s, that left mainly the old South East factories still using the wet process and the solution was to construct a giant new wet process factory in Kent – Northfleet Works near Gravesend – which would produce 3.6 million tonnes a year of cement or six to seven times as much as any Blue Circle cement works had done before, thereby achieving economies from this new factory; and then close the smaller and older factories. Northfleet was duly opened 1970-71 and large numbers of old works closed.

However, Northfleet was designed to manufacture ordinary portland cement, the "base load" of the industry and a few older works were kept in operation, on a reduced scale, to make various specialist types of cement. Shipton closed two of its three kilns in 1975, but the third was retained to manufacture "Packbind", a type of cement in demand in the coal mining industry, at a rate of around 90,000 tonnes a year. As this continued and as problems (notably huge increases in energy prices) began to beset Northfleet, it was noted that Shipton was by far the nearest cement works to London and the South East to have access to dry raw materials which could be used in a dry cement process. That could give Blue Circle a tremendous advantage if a new dry process plant could be built on the Shipton site, which was less than 50 miles from London. Accordingly, studies and designs began and in 1983 a planning application was prepared to build a new dry process plant that would produce 750,000 tonnes of cement annually, at a far lower cost than the Northfleet plant. The new plant would be located immediately west of the existing Shipton kiln and replace it, but the existing cement milling, bagging and loadout facilities would be retained and revamped.

At that time, the Shipton company was working under the Town and Country Planning Act of 1947 and the new quarry workings were to be conducted on lands to the west of the factory. However, in 1946 negotiations were held with the Oxfordshire County Council with a view to improving consented areas under the Interim Joint Working Scheme. The applications of 1947 included not only the land within the existing workings but also fields 7, 10 and 18 in Bunkers Hill area and fields 2, 5 and 8 further to the west at Weaveley Furze.

In the event the application to work these two areas i.e. Bunkers Hill and Weaveley was withdrawn on the understanding that Oxfordshire County Council would not pursue any other uses of these lands for a period of three years. On the 30th June 1948 permission was again granted to extract stone for the purpose of portland cement manufacture from the existing Shipton quarry. At that time the depth of working was expected to be 50ft and the reserves estimated at 32 years. In 1950 another application was made to work the above-mentioned fields. As the new Oxfordshire County Map was being prepared at the time, these fields were then marked 'for mineral working.'

Again in 1956 Blue Circle had approval to extract materials to be used in portland stone manufacture from these fields. The County Council also approved fields that were not in the ownership at that time by the Blue Circle company. Permission was not granted until 2nd of February 1958 with a strange condition – *that the holly hedge be kept to a height of at least 5 ft.* Because mineral extraction had not commenced at the time, extensive replacement of the hedge was carried out in 1966.

By 1977 further approval was given to quarry at lands adjacent at nearby Manor Farm.

In March 1979, two small excavations were dug in the area that had previously been out of limits although without Oxfordshire County Council taking any enforcement action.

During 1978 and 1979 several discussions were held with Oxfordshire County planners concerning the possible opening of land at Bunkers Hill for future quarrying. There were various planning limitations from the 1956 planning permission as regards what land could be used. Negotiations between Blue Circle and

Plate 64: Compare this 1970s view with the image in Plate 63 taken some 30 years previously. It shows how the quarry has increased in size and how near it has become to the Bunkers Hill houses (bottom left). Several of the internal railway tracks can clearly be seen. Courtesy The Blue Circle Magazine, Vol 26, No. 2, Winter 1972.

the Council were to be resolved to the Council's satisfaction upon the following matters: the submission by the company of a restoration scheme for the land prior to the commencement of development; the hours of working to be permitted and the need for the imposition of a limit on the production capacity.

Blue Circle then, in August 1980, had discussions with the County Planning Officer and the Minerals Officer on revised plans for the Shipton works and the company's plans for modernisation or re-building of the plant. In February 1982, this was a prelude that the company wanted to go forward with submitting a detailed planning application, entitled 'THE NEW WORKS' and more quarrying in the area, although this was eventually delayed until 1983.

There was considerable correspondence between Blue Circle, Oxfordshire County Council and the Ministry of Works regarding the nearby Bunkers Hill Long Barrow, situated on Blue Circle land. As stated in a previous letter, questions were sent asking to hear about any steps required regarding the Long Barrow removal, particularly allowing the Ministry to remove any remains before it was destroyed. An answer from the Oxfordshire County Council said that the Mound (Bunkers Hill Long Barrow) must be scientifically examined by the Ministry of Works at least one to two years prior to its destruction.

Correspondence from the Ministry of Works stated that there was some doubt as to the existence of the Long Barrow, though they still wanted to investigate it before it was destroyed. Although the following has been mentioned before it is included again here in the context of the planning proposals.

The Royal Commission on Historical Monuments was consulted in 1981 and they only had on their records regarding Bunkers Hill Long Barrow was that it was of 70 metres in length x 17 metres in width by 0.3 metres in height. Its axis was 15 degrees of true north and that appeared to have no side ditches and that there was no evidence to suggest that it necessarily still existed. The entire area had been walked in search of it and that they were convinced that there was nothing there, implying that it had probably been ploughed in over the years. The barrow had not been archaeologically investigated by the Commission but there was a possibility that the Oxfordshire Archaeological Unit may have? It was suggested that they may be contacted – the investigation then disappears from the files!

One disturbing fact recorded in the following proposals on these 'new works' was that the houses at Bunkers Hill were to be destroyed in favour of the Shipton quarry being extended and a new road constructed to run round this area, so it could be quarried. There was no mention in the paperwork of when or where the inhabitants of Bunkers Hill would go or what new accommodation would be built for them. This is called the age of progress!

This important 1982 planning application was very controversial but appeared at the time likely to be successful.

The Blue Circle company even went public and the accompanying brochure shows its public consultation events.

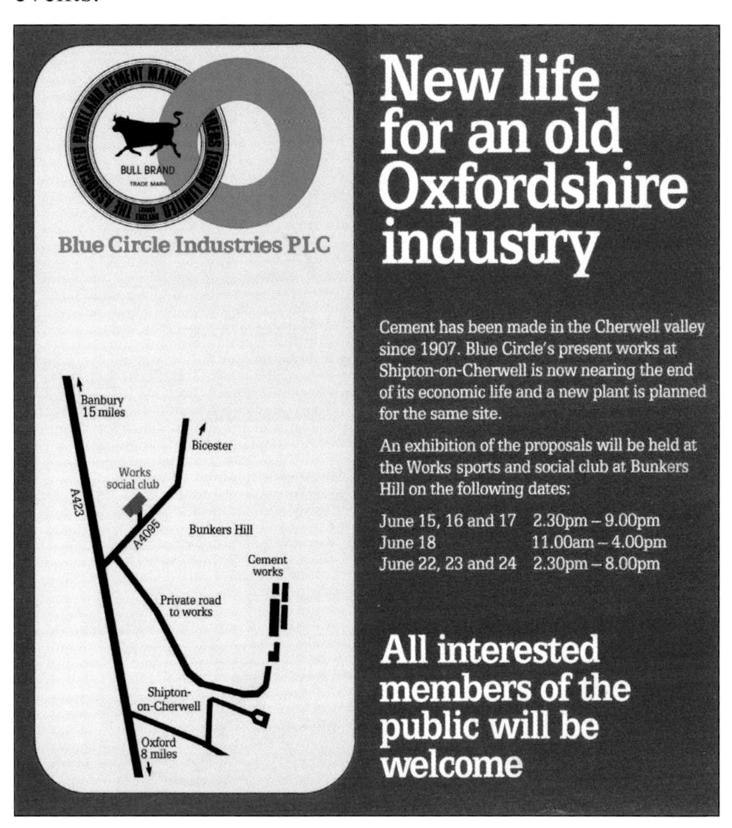

The company had drawn up numerous development statements for this planning application and re-entitled them all 'THE IMPORTANT NEW SHIPTON WORKS'!

Amongst the reams of paper for this application were some interesting ideas and proposals and some are listed here. Because the proposed new works were going to be larger and more efficient and the demand for cement was estimated to increase, one of the great concerns were the methods of despatch from Shipton works.

These were mainly driven by cost, the effects of vehicle movements on the local roads (suggestions were made of a completely new local road being constructed around the works area) plus a large increase in rail usage to, which British Rail initially answered that they could not handle anything above the existing contract.

The rail problem was one of the big stumbling blocks, as the equipment proposed to be installed at the rail sidings at Shipton would load 20 wagon trains in two hours. At the new works, wagons would be loaded four at a time, necessitating five individual train movements during loading.

This could involve a British Rail locomotive, standing and operating on Blue Circle's sidings during the loading period. If this was not possible, then Blue Circle would buy a suitably powered locomotive for this purpose. The loaded trains would then stand with a British Rail locomotive at its head and be absorbed into the existing timetable. This was all speculative, however, as British Rail had already declined to invest any money and claimed not to have spare paths for the proposed traffic and insufficient locomotives to work them. One other factor was that rail costs at this time were increasing disproportionately compared to road costs.

Regarding road usage, at the present time vehicle despatches would normally (due to planning restrictions) take place between 6am and 4pm but large construction projects coming on line at that time in the UK would need lorry departures outside these times. Blue Circle had heard from parliamentary sources that the maximum permitted road-vehicle size was likely to increase during the life of the new works and so they could take advantage of this, but this would involve a larger investment in lorries especially with the BR's negative answer. There were questions about how many lorries would they need to compete with one 20 wagon train load of around 750 tons, that could be unloaded in about four hours. There was also the cost of a 20-wagon train at approximately £700,000 at that time.

Regarding the buildings on this new works; in the proposed design all the new buildings would be of a low modular style incorporating many of the existing individual buildings, into just one building of rectangular form, composed of smooth modular components within an adaptable cladding system. Machinery required to be housed to exclude weather and to comply with current noise and dust emissions limits. Screening by extension of the silo walls and the fuel and raw materials were to be low domed forms. All the site machinery was to be controlled from one main control centre, which would improve safety of the staff and site.

It went on to state that "at present the works are totally dependent upon the operation of the quarry." That indicated there was virtually no 'buffer' storage of quarried raw materials ready for use. In the case of the proposed new works, blended raw material for up to seven days kiln operation would be kept as a buffer between the works and the quarry operation, which could imply night time work in the quarry?

Within these pages are also actual and projected lists of the lorry movements at the works; firstly actual averages by road during the period 1977 to 1981:

Materials exported included:

Limestone to Dunstable Works – 45 miles East – 12,631 tons per year – 4 daily journeys

Limestone to Masons Works – 113 miles East – 2,193 tons per year – 3 daily journeys

Limestone to Westbury Works – 69 miles West – 1,656 tons per year – 3 daily journeys

Materials imported:

Clinker ex Martin Earles – South – 30,695 tons over a 9 months period – 52 journeys per week by a hired haulier.

Clinker ex Wouldham – South – 2,273 tons over a 9-week period – 17 journeys per week by a hired haulier.

Clinker ex Cauldon Works, Staffordshire, for grinding at Shipton – 13,230 tons

Clinker ex Barnstone Works, Nottinghamshire for 'PACKBIND' – 7,049 tons

Gypsum from Kilvington Quarry, Nottinghamshire, for use at Shipton – 6,545 tons

Then in comparison, the following *imported tonnage of materials* handled by BR averaged during this same five year period was:

	Clinker ex Martin Earles Works, Kent	1,928 tons
	Cement ex Metropolitan Works, Essex	6,699 tons
	Steam coal	200 tons
	Kiln coal	68,855 tons
	Gypsum	17,584 tons
Cement despatched	Ordinary portland cement (bags)	1,795 tons
	Ordinary portland cement (bulk)	20,312 tons
	Walcrete (a special cement) bags	98 tons

During this period, the factory at Shipton was reduced to just one-kiln operation and over this same 1977-1981 period, the average annual cement tonnage despatched by road was 105,040 tons in bulk and 17,170 tons in bags, all of which were 'Packbind' for the National Coal Board.

As can be seen, Shipton-on-Cherwell works was heavily involved in both sending raw materials to other Blue Circle works and receiving cement clinker from other factories for grinding into cement.

So finally, since the reduction to one operating kiln in 1975, the only rail traffic remaining was coal from Daw Mill Colliery in Warwickshire. Overall, the works was then almost wholly reliant on road transport.

After the last kiln closed in 1986, the grinding mills and downstream plant continued to operate to produce 'Packbind' plus continuing to supply the local cement market until customers could become accustomed to purchasing cement delivered direct from elsewhere.

Unfortunately, Blue Circle was becoming disenchanted with the returns from cement manufacture and were looking more towards dis-investment than reinvestment. This huge planning application (which had involved a high investment and been very time consuming) was withdrawn before Council determination and instead the main Shipton-on-Cherwell cement works closed in 1986. So far as it is known, no rail traffic was involved in either direction after this date and cessation of clinker grinding and final complete closure of the Shipton site seems to have been in 1993.

It would be appropriate to record the managers of the Shipton Cement works, as known, and record their service, not only to the Shipton-on-Cherwell works, but to the cement industry in general. So far as it is known, they were:

c1927	–	1930	H.P. Lidgey
c1931	–	1935?	C.E. Dunleavie
1936	–	1939	W. Swan
c1941	–	1944	Colin Tetley
1945	–	1968	Leonard Jupp (one of longest serving managers at any one Blue Circle location).
1973	–	1974	Jim McColgan
1975	–	1983	Ron Wharam
1984	–	1986	Keith Maltas
1986	–	1993.	Raymond Jupp (son of Leonard Jupp)

Chapter Seven

SHIPTON-on-CHERWELL QUARRY FLOOR MACHINERY and EQUIPMENT.

The first Ruston* steam excavator, a Ruston 25, appeared at Shipton-on-Cherwell quarry in 1928, its purchase being reported at the fifth Annual General Meeting of the company.

Plate 65: First, a rather poor photograph taken around the early 1930s, shows the quarry gang – the men who worked the quarry face, standing in front of the Ruston 25 steam excavator with locomotive ALBION No. 1 (AB 776/1886) on the left with some internal quarry wagons attached.

Courtesy; from the 50-year Alpha cement booklet from the collection of John Oxford.

It is therefore appropriate that the all-important excavators and machinery are dealt with at this stage in the history of the Shipton cement works. The plant could not have worked to its capacity if these huge machines had not been around and, as Shipton had several during its working life, these will be dealt with in detail in this chapter, together with some of the other machines that were used at later dates in the quarry's life.

The first steam powered excavator machine constructed by Ruston in the 1920s was designated Ruston 25 No. 1, which was specifically designed for quarry work and was supplied to the Oxford & Shipton Cement Ltd. in 1928 *(see the Chairman's address for that year)*. The whole machine weighed 112 tons, supported on a standard Ruston underframe with two crawler tracks.

* Ruston, Proctor & Co. Ltd. and its successor, Ruston & Hornsby Ltd., fulfilled most of the orders in the market at that time, with Ruston & Hornsby Ltd. having been formed when Ruston Proctor & Co. Ltd. joined with Richard Hornsby & Sons of Grantham.

Plate 66: The first Ruston 25 crawler mounted quarry excavator when new and on test before delivery to Shipton quarry. Note that much of the machines cladding has yet to be fitted.

Author's collection acquired from an international auction site.

Plate 67: A fine view of the boiler, feed water pump, engine and winding gear of the Ruston 25 with only the rear section of cladding in place. Note to the wheels on which the whole bodywork and machinery rotates.

Author's collection acquired from an international auction site.

Plate 68: The Ruston 25 excavator photographed at work on the Shipton quarry face in the1930s. Many other cement works were modernised in the 1925-35 period and most of their quarries were provided with modern electric face shovels and not steam. Note the bags of coal under the back of the excavator.
Courtesy Oxfordshire County Council, Oxford History Centre; ref POX0100229.

There were eight Ruston 25s built and after Oxford's No. 1 machine, the next four were also sold in Britain. The last three finished were built under the name Ruston-Bucyrus, after the American excavator company Bucyrus-Erie, who had formed a joint company with Ruston & Hornsby in 1930.

By the 1930s however, the end was in sight for the steam powered excavators in general, but for one lone steam operated Ruston-Bucyrus 52-B, works number 2686, which was delivered to the Alpha Cement Co, Oxfordshire, in March 1935, for work at Shipton-on-Cherwell quarry.

The new excavator was smaller having a 2¼ cubic yard bucket and reportedly weighed a total of 84 tons. The machine was equipped with a locomotive style boiler to supply the three individual steam engines and had a large water tank, all of which also acted as counterweights to the boom weight. The main twin cylinder hoisting engine also drove the crawler tracks via clutches. One of the two, smaller, twin cylinder engines was used both to raise and lower the boom and to drive the rotation of the machine; its twin, which was mounted on the boom, drove the racking gear.

Steam was circulated by means of piston valves and the operation of the machine was controlled by friction style clutches and brakes, with steering clutches manually operated from the ground position.

The enclosed hoisting steam engine was fitted with ball and roller bearings, which showed the modernisation that had progressed within the industry and so justifying the statement in the brochure that this machine was 'state of the art.'

Plate 69: The 52-B standing in the Shipton quarry in 1969 with the quarry face and some of the quarry rail tracks visible. This view shows well the two crawler tracks supporting the machine and the racking gear steam engine mounted on the beam. *Courtesy Keith Haddock collection.*

Plate 70: The compact machinery of the 52-B, No. 2686 seen before leaving the Ruston works.
Author's collection reproduced under the Creative Common licence.

Plate 71: This photograph shows the 52-B steam shovel machine No. 2686 at Shipton cement works quarry towards the end of its working life in the late 1960s, looking rather tired and worn. *Courtesy of the Leicester Museum.*

Plate 72: No more steam excavators were purchased for Shipton quarry and, marking their end, this is a view of 52-B No. 2686 shortly after being taken out of use. *Courtesy Kevin Lane collection.*

Plate 73: Following being taken out of use at the Shipton cement works quarry, the 52-B No. 2686 is being made ready for its dismantling and removal in the early 1970s to the East Midlands Museum of Technology at Leicester. Courtesy of the Leicester Museum.

The excavator was reportedly to have performed extremely well both in removing the top overburden and excavating the all-important limestone. The excavator's original operator (when interviewed some years later) said about machine No. 2686, that the ease in removing the heavy rock and the speed of operation in general was due to the smooth slewing and dumping cycle. The joke was, that in the first few days of operation, the second man employed on the machine acting as the boiler stoker, tending the steam boiler at the rear was continually being thrown off the back of the machine by the fast-slewing motion of the machine. He was eventually equipped with a harness and tethered to the footplate to avoid any further mishaps!

No. 2686 lasted at Shipton until about 1970, when expensive boiler repairs were required and old age resulted in the company deciding to retire it.

Around that time a letter was written on behalf of the Alpha Cement Co, offering to donate No. 2686 to the East Midlands Museum of Technology. Later, in 1971, the machine was dismantled by the Industrial Steam Preservation Group ready for Pickford's transportation to the Museum. It is now on public display after restoration work.

Although ordered in 1938, it is not certain exactly when the next excavator came on-line. It was a large Ransomes & Rapier model 5160 electric machine, built at their Ipswich factory. In *Figure 38* it shows the information box section of the 3ft x 2ft 6in general arrangement drawing dated June 1939 and with the customer as Alpha Cement Ltd. This indicates that Shipton had two excavators in use by the early 1940s. Due to the size of this monster machine, it would have taken many months, possibly a year or more, following its delivery in March 1939 to complete its erection, with components being delivered by road from the works.

Model 5160 was classified as a stripping shovel and unusually had four crawlers supporting its bulk, as seen in *Plates 75, 77 and Figure 37.*

Plate 74: The Shipton cement works third purchase was a much larger excavator and this wonderful image is of the Ransomes & Rapier model electric 5160 that was built in 1938, loading stone into a train in the charge of No. 3, a 0-6-0ST built by Andrew Barclay (AB 2041/1937). The quarry face and the various levels of materials is well illustrated in this view! *Courtesy John Scholes collection.*

Only two of these huge machines were ever constructed by Ransomes & Rapier – the Shipton machine was the second; the first, built in 1936, worked in the ironstone quarries based in the Midlands.

The boom length was 57ft and the bucket 3½ cubic yards and this machine was equipped with hydraulic levelling devices on the crawler assemblies. The overall weight was 134 tons 14 cwt.

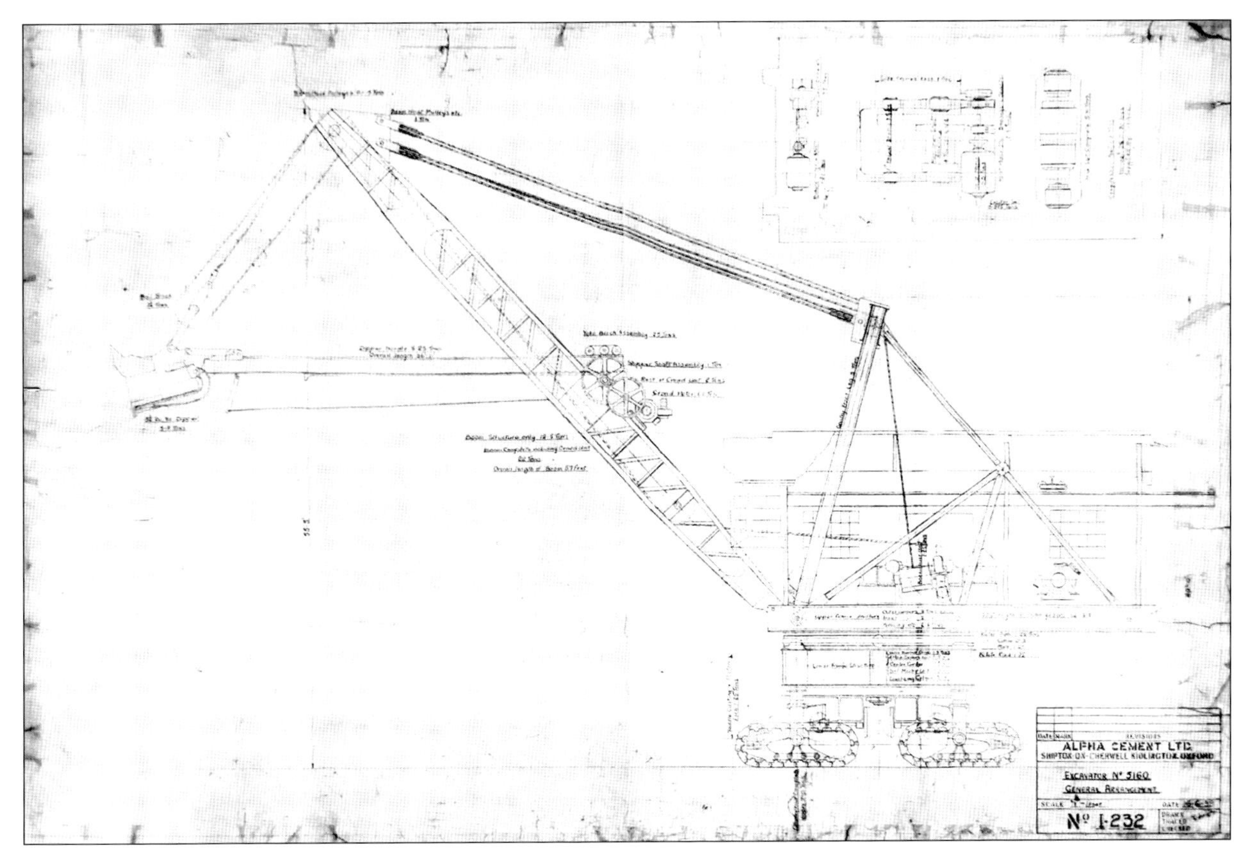

Figure 37: This is the best reproduction obtainable from a folded copy of a very large General Arrangement drawing of the 5160 excavator. *Courtesy Keith Haddock collection.*

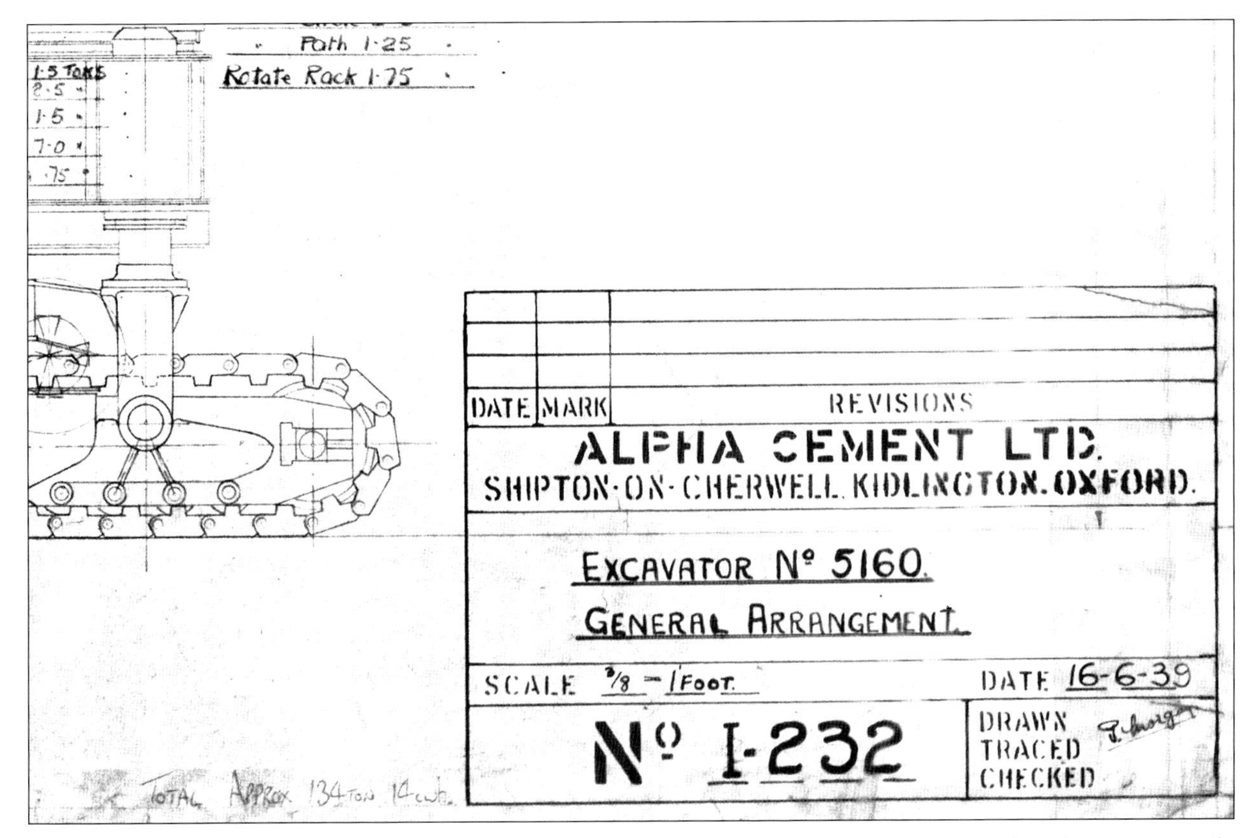

Figure 38: An enlargement of part of the drawing in Figure 37, showing the customer's details. Interestingly, the machine had been delivered and was being erected by the time this drawing was made – was it to show the final 'as built' configuration? *Courtesy Keith Haddock collection.*

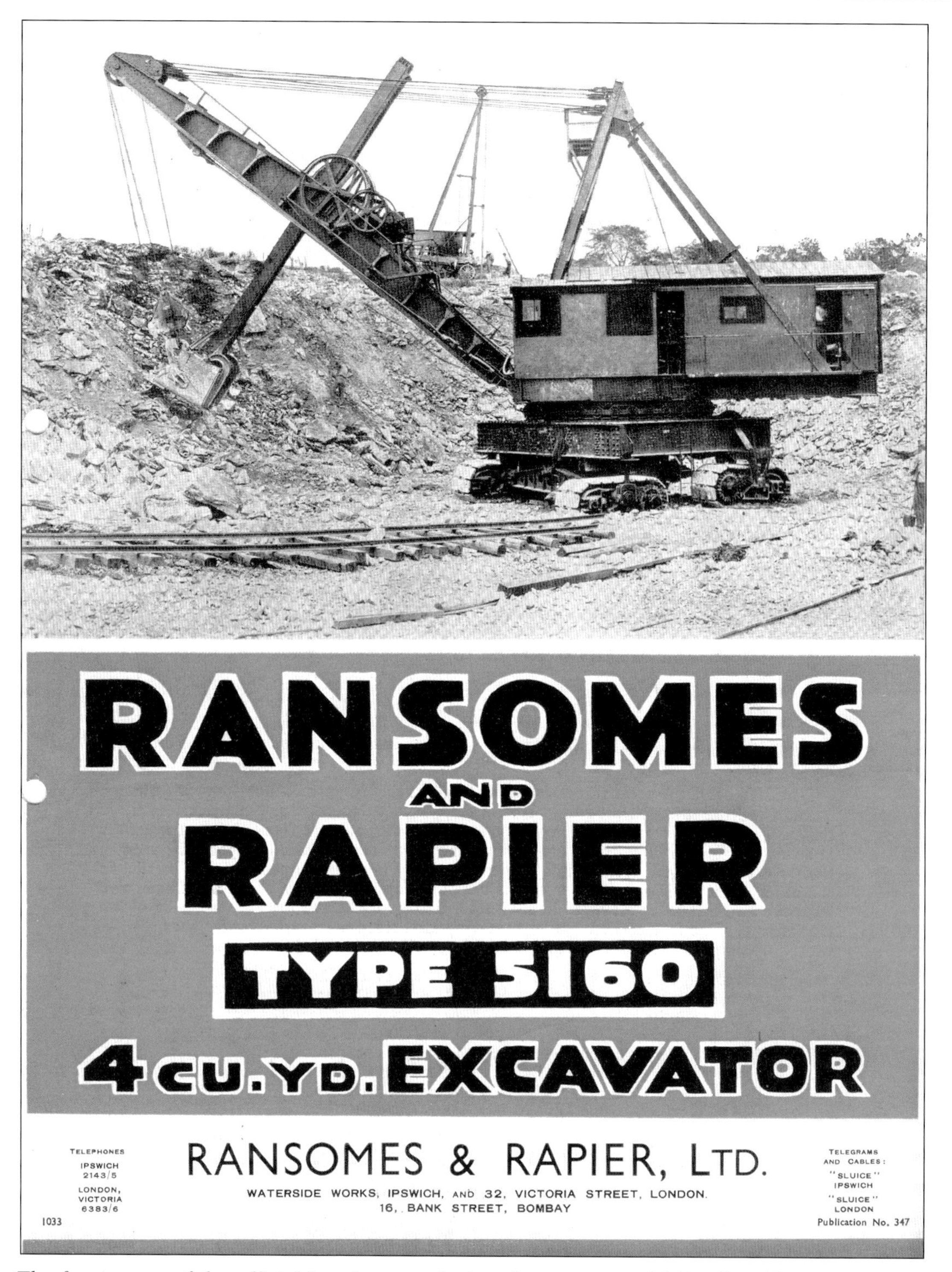

Plate 75: The front cover of the official brochure marketing the excavator 5160 offered by Ransomes & Rapier of Ipswich. *Courtesy Keith Haddock collection.*

This sales brochure showed a different machine to that of the actual Shipton machine. This was because Ransomes & Rapier started building excavators in 1926 under licence from Marion Steam Shovel Co in the USA, however the specification in this brochure describes a Rapier machine.

The picture is of an early US built Marion model 5160. This has a three-point suspension, a levelling beam across one end linking two of the crawler assemblies, whereas the British built machine supplied to Shipton had a vertical hydraulic cylinder linking each crawler to the superstructure. One of these may be seen in *Figure 38*.

Working Ranges for Type 5160—with Shovel Equipment.

Boom at angle of 45 degrees.

A—Length of boom … … … … … …	35′ 0″	45′ 0″	55′ 0″
(metres)	10.668	13.716	16.764
B—Length of dipper handle … … …	25′ 0″	30′ 0″	35′ 0″
(metres)	7.620	9.144	10.668
C—Size of dipper … … … … … …	4 cu. yd.	3 cu. yd.	2½ cu. yd.
(cu. metres)	3.06	2.295	1.912
D—Height of dump (handle horizontal) …	21′ 2″	25′ 6″	30′ 0″
(metres)	6.451	7.792	9.144
E—Height of dump (maximum) … …	27′ 5″	35′ 8″	43′ 5″
(metres)	8.356	10.871	13.333
F—Height of boom … … … … … …	36′ 11″	44′ 0″	51′ 0″
(metres)	11.252	13.411	15.545
G—Cut below grade … … … … …	3′ 0″	3′ 0″	3′ 0″
(metres)	.914	.914	.914
H—Radius of boom … … … … … …	35′ 3″	42′ 4″	49′ 5″
(metres)	10.774	12.903	15.062
J—Radius of dump … … … … … …	45′ 0″	52′ 6″	59′ 6″
(metres)	13.716	16.002	18.135
K—Radius of dump (handle horizontal)	45′ 7″	54′ 0″	62′ 0″
(metres)	13.893	21.26	24.409
L—Radius of clean up … … … … …	30′ 8″	35′ 0″	38′ 6″
(metres)	9.347	10.668	11.734
M—Centre of rotation to front cab … …	8′ 6″	8′ 6″	8′ 6″
(metres)	2.59	2.59	2.59
N—Centre to centre of crawler trucks …	14′ 0″	14′ 0″	14′ 0″
(metres)	4.267	4.267	4.267
P—Overall length of crawlers … … …	22′ 10″	22′ 10″	22′ 10″
(metres)	6.755	6.755	6.755
R—Height of gantry … … … … … …	36′ 0″	36′ 0″	36′ 0″
(metres)	10.973	10.973	10.973
S—Radius to corner … … … … … …	22′ 0″	22′ 0″	22′ 0″
(metres)	6.705	6.705	6.705
Approximate shipping weight (less ballast) … …	145 tons. 147.320 kgs.		
Approxinate working weight … … … …	158 tons. 160.400 kgs.		
Ballast for upper frame (furnished by buyer) …	22 tons. 22.200 kgs.		

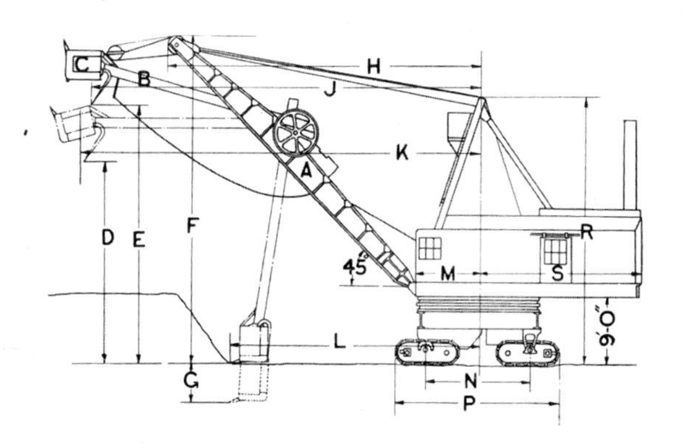

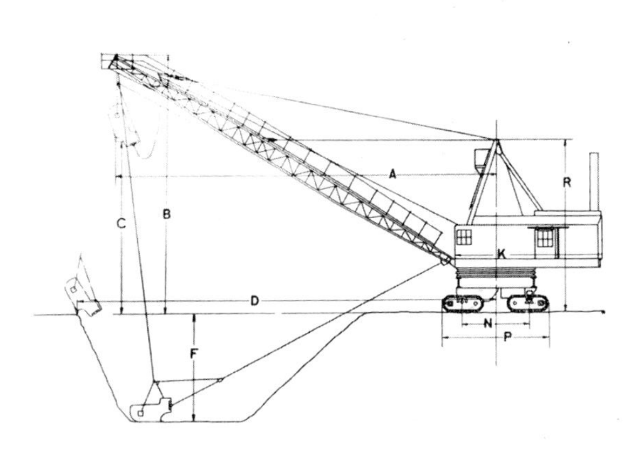

Working Ranges for Type 5160—with Dragline Equipment.

Boom at angle of 30 degrees.

Length of boom … … … … … …	80′ 0″	90′ 0″	100′ 0″
(metres)	24.384	27.432	30.480
Size of bucket … … … … … …	4 cu. yd.	3 cu. yd.	2½ cu. yd.
(cu. metres)	3.06	2.295	1.912
A—Radius of dump … … … … … …	80′ 6″	89′ 3″	97′ 9″
(metres)	24.536	27.204	29.794
B—Height of boom … … … … … …	53′ 0″	58′ 0″	63′ 0″
(metres)	16.154	17.678	19.202
C—Height of dump … … … … … …	30′ 9″	36′ 6″	43′ 3″
(metres)	9.372	11.125	13.[illegible]
D—Digging radius at grade … … …	95′ 0″	107′ 0″	120[illegible]
(metres)	28.956	32.614	36.576
*F—Depth of cut … … … … … …	45′ 6″	40′ 3″	34′ 0″
(metres)	13.868	12.268	10.363
R—Height of gantry … … … … …	36′ 0″	36′ 0″	36′ 0″
(metres)	10.973	10.973	10.973
K—Length of upper frame … … … …	30′ 6″	30′ 6″	30′ 6″
(metres)	9.296	9.296	9.296
N—Centre to centre of crawler trucks …	14′ 0″	14′ 0″	14′ 0″
(metres)	4.267	4.267	4.267
P—Overall length of crawlers … … …	22′ 10″	22′ 10″	22′ 10″
(metres)	6.755	6.755	6.755
Approximate shipping weight (less ballast) …	138 tons. 140.000 kgs.		
Approximate working weight … … … …	155 tons. 157.200 kgs.		
Ballast for upper frame (furnished by buyer) …	24 tons. 24.300 kgs.		

*Calculated for one layer of rope. With 2 layers depth increased according to conditions.

TELEPHONES: IPSWICH 2143/5 LONDON, VICTORIA 6383/6

RANSOMES & RAPIER, LTD.

WATERSIDE WORKS, IPSWICH, AND 32, VICTORIA STREET, LONDON.
16, BANK STREET, BOMBAY.

TELEGRAMS AND CABLES "SLUICE" IPSWICH "SLUICE" LONDON

1033

Publication No. 347

Figure 39: The specification details of the 5160 excavators from the sales brochure. It was not adapted to dragline mode at Shipton. Courtesy Keith Haddock collection.

Plate 76: The electric 5160-excavator machine at work in the Shipton quarry in 1969 and the electric power cable can clearly be seen lying next to the diesel-powered compressor? Note the usual five rake of wagons being loaded. Courtesy photographer Keith Haddock.

Plate 77: The huge and complicated four crawler assemblies that supported this enormous monster excavator. Note the steps on the left to get up into the control cab some 21 feet in the air! Courtesy photographer Keith Haddock.

The photograph in *Plate 76* was taken in 1969 and this was the last report of the R&R 5160 in use. With the last steam excavator also being taken out of service at this time, the quarry became reliant on another electric shovel, a Ruston-Bucyrus RB120. Between 1948 and 1954, some 30 of this model were built with APCM (Blue Circle) purchasing six, one each in 1951 and 1952, three in 1953 and the last one in 1954. It is not known which one of the six went to Shipton.

The RB120 had a 5 cubic yard bucket and weighed 162½ tons. The quarry face cutting height was approximately 33 feet and the boom swing from face to the receiving trucks could be up to 44 feet.

In 1975, Shipton works reduced capacity from three kilns to one, meaning that quarry output fell to about a third; so, the RB120 would have been adequate for the remaining work. In the proposal for a 'new works', dating 1981-82, the stated intention was to change to a 3-bench system, and use three hydraulic shovels, each with a bucket capacity of about 3 cubic metres. Two machines would work on benches 1 and 2 and, presumably, would be face shovels. The third would work bench 3, but from a position on bench 2, so would have been a 'back hoe' design. Existing quarry equipment was also listed in the proposal, including the RB120.

Clearly the huge R&R 5160 was no longer in use or even present by the 1981-82 era so that, with the quarry still on its 2-bench configuration, the RB 120 would have stood on the top of bench 2 to dig out bench 1 and that the O&K RH9 excavator would have been stood on bench 2 to dig bench 2. This however is all partly guess work!

Plate 78: An RB120 at work in an unknown quarry. Note the electric power cable running to the excavator.
Courtesy Keith Haddock collection.

Also listed is an O&K RH9 hydraulic face excavator and loader, a versatile machine which could be used on a number of tasks. It is thought that, under the existing 2 bench system, the RB120 would have worked bench 1 from a position on bench 2 and the OK RH9 would have stood on bench 2 and worked that bench. This, however is merely conjecture on part of the author!

Plate 79: The other side of the versatile electric RB120, again at another quarry almost certainly in the USA as it is a Bucyrus-Erie built machine but showing the loading technique into a dumper truck.
Courtesy Keith Haddock collection.

Plate 80: A good side view of the smaller more up-to-date O&K RH9 hydraulic face excavator and loader at an unknown quarry.
Authors collection obtained from an international auction site and reproduction under the Creative Common licence.

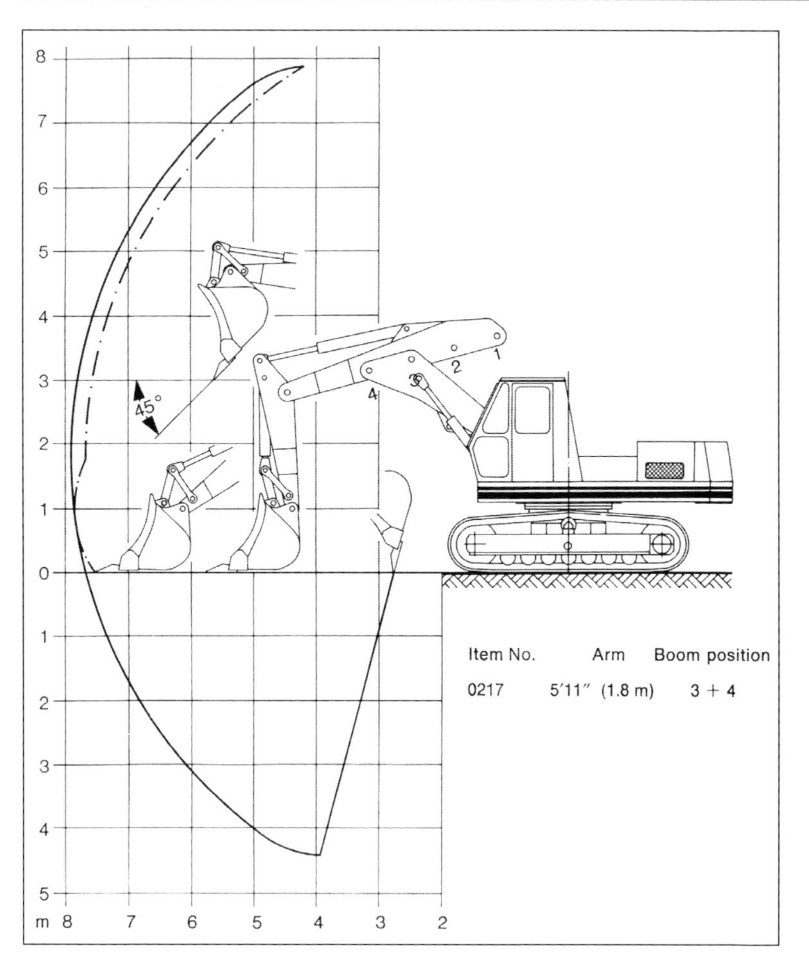

Figure 40: Showing the large operating range available with this type of machine.
Courtesy Keith Haddock collection.

Two more machines are listed; an O&K MH6 wheeled excavator, as shown in *Plate 81*, but fitted at Shipton with a hydraulic breaker rather than a bucket. This was used to deal with oversized rocks. Finally, the quarry had a Caterpillar 988 wheeled front end loader.

Plate 81: The even more modern O&K MH6 four-wheeled excavator was highly manoeuvrable and was used to break large rocks at Shipton. This view is of a standard machine, fitted with a 'back hoe' bucket.
Courtesy Keith Haddock collection.

Plate 82: These CAT 988 loaders were a very good workhorse for general movement of the stone and other materials in and around the quarry site and replaced some of the rail operations within the quarry.
Courtesy Keith Haddock collection.

The 1980 inventory stated that 3 x 28 tonne Aveling-Barford 'Centaur' dump trucks were in use. These very versatile vehicles were made at the Aveling-Barford works at Grantham.

Plate 83: Probably the most up-to-date pieces of equipment that Blue Circle purchased before the final closure of the Shipton works and quarry. With the amount they held, the manoeuvrability and cost to operate they would have been the ultimate in moving great loads to the crushers. *Courtesy Keith Haddock collection.*

Over the years there were probably many other machines used in the quarry. These were not itemised in the 1980 inventory and are therefore omitted.

CHAPTER EIGHT

The LOCOMOTIVES, WAGONS etc., of the OXFORD & SHIPTON CEMENT LTD, ALPHA CEMENT LTD, APCM (BLUE CIRCLE) and BLUE CIRCLE INDUSTRIES PLC.

LOCOMOTIVES No. 1 AND No. 2 ATKINSON-WALKER – WORKS NUMBERS 109 AND 110

Atkinson-Walker Wagons (or Waggons as both spellings used) Ltd. was created by a partnership between Atkinson Vehicles Ltd. (who had, in November 1926, purchased the steam wagon stock from Leyland Motors Company at Chorley) and was based at Frenchwood Works, Preston and Walker Brothers (Wigan) Ltd., Pagefield Ironworks, Wigan, that had been founded in 1866.

The two companies joined forces in late 1925. Atkinson's main business was the construction of road vehicles and Walker Brothers was experienced in railway engineering, having, in the period 1872-1883, produced around twenty-five shunting locomotives, mainly for local collieries.

So, just as Oxford & Shipton Cement Ltd. was looking for locomotives for its new works, this novel design was being promoted – perhaps with attractive prices?

The "Atkinson-Walker" Rail Tractor

ANY RAIL GAUGE FROM TWO FEET UPWARDS.

THE ATKINSON WALKER RAIL TRACTOR

ANY POWER TO 1,000 TONS ON LEVEL.

Prices and full particulars from :

ATKINSON WALKER WAGONS, LTD.,
PRESTON, ENGLAND.

TELEGRAMS: "WAGONS, PRESTON." CODE: BENTLEY'S. TELEPHONES: PRESTON 2131 AND 2132.

Plate 84: The Atkinson-Walker advert for the four-wheel version of their Rail Tractor.
Courtesy Industrial Railway Society.

During the 1930s, this combination of the two companies constructed some diesel railcars for the 3-foot gauge West Clare Railway in Ireland plus a few heavy road cranes, some owned by the LMS.

Geared steam railway locomotive production only seemed to last just three years (1927 to 1930) and the table below lists the known sales dates and basic details. Sadly, all the company's records and drawings were destroyed by Atkinson Lorries (1933) Ltd., later part of Seddon Atkinson.

These geared steam locomotives only seemed to be popular for about 15 years from the cessation of WW1, when the new diesel locomotives became available with the notable exception of those produced by the Sentinel Wagon Works of Shrewsbury, which continued to manufacture the type until 1958. By the 1930s their place in the market was being filled more and more by diesel locomotives.

Atkinson-Walker Wagons Ltd., locomotives built during 1927 to 1930			
Locomotive classes and details. The date shown is probably date of sale and not of construction.			
A	Four 3 ft wheels, outside bearings, 2 – cylinder vertical steam boiler engine with 7" x 10" cylinders		
B	Same as class A but with three cylinders		
C	Four 2 ft 3 ins wheels, inside bearings, 2 – cylinder horizontal steam engine with 7" x 10" cylinders		
D	Six 3 ft wheels, outside bearings, 4 – cylinder vertical steam engine with 7" x 10" cylinders		
NG	Four 2 ft 4 ins wheels, outside bearings, vertical cylinder steam engine, various rail gauges		
Works no	***Date***	***Class***	***Description***
101	c1926	prototype?	Singapore Municipal Council?
102	1927	A	New to Henry Leetham & Sons Ltd. Flour Mills, York. Scr 1948?
103	1927/28	B7	New to Hutchinson Hollingsworth & Co. Ltd.? but later to Walker Bros., Wigan for works shunting, named MARY. Rebuilt in 1947 with a diesel engine
104	1928	B	New to Richard Briggs & Sons Ltd., Clitheroe named LAZARUS
105	1928	NG	To Singapore Municipal Council as No. 1
106	1928	NG	As above but numbered No. 2
107	1928	NG	As above but numbered No. 3
108	1928	NG	As above but numbered No. 4
109	1927	A	New to the Oxford & Shipton Cement Limited as No. 2
110	1928	D	New to the Oxford & Shipton Cement Limited as No. 1; later named GOLIATH when at Alpha Cement Ltd., West Thurrock in 1940
111	1928	A	3 ft gauge to Ivybridge China Clay Co; named LADY MALLABY-DEELEY
112	1928	B	New to Shap Granite Co. Ltd.; named FELSPAR 2
113	1928?	C	New to Blaxter's Quarries Ltd., Elsdon, Northumberland. In May 1947 working later at Tottenham & District Gas Co.
114	1928	A3	3 ft gauge to Clogher Valley Railway but refused and returned
115	1928	A1	No details
116	1928	A1	No details
117	1930	B	New to H. Arnold & Sons, contractors Doncaster as No. 1
118	1930	B	New to H. Arnold & Sons, contractors Doncaster as No. 2; used by the LNER
119	1930	B	New to H. Arnold & Sons, contractors Doncaster as No. 3. used by the LNER
Odd No. 31790	?	?	To the Shap Granite Co. Ltd., named WASDALE – scrapped. c1952

The above basic list has been prepared from the impressive history of these vertical boiler locomotives and only shows the first sale to the customer. The book **Vertical Boiler Locomotives and Railmotors built in Great Britain – Volume Two** by Philip Ashforth and Vic Bradley published by the Industrial Locomotive Society in 2016, contains a full, comprehensive detailed history of the company and of each locomotive.

The two Atkinson-Walker steam vertical boiler locomotives were the first two purchased by the Oxford & Shipton Cement Limited and unfortunately seemed to be a very bad choice. Firstly, these builders had had little experience in industrial style motive power and had no previous track record. So, the two locomotives were purchased in 1928 but, by the time of the new Alpha Company Ltd. take-over in July1934, they were deemed unsuitable and were made redundant (both during 1937) being replaced by more suitable conventional locomotives from established industrial locomotive builders, as can be seen later in this chapter.

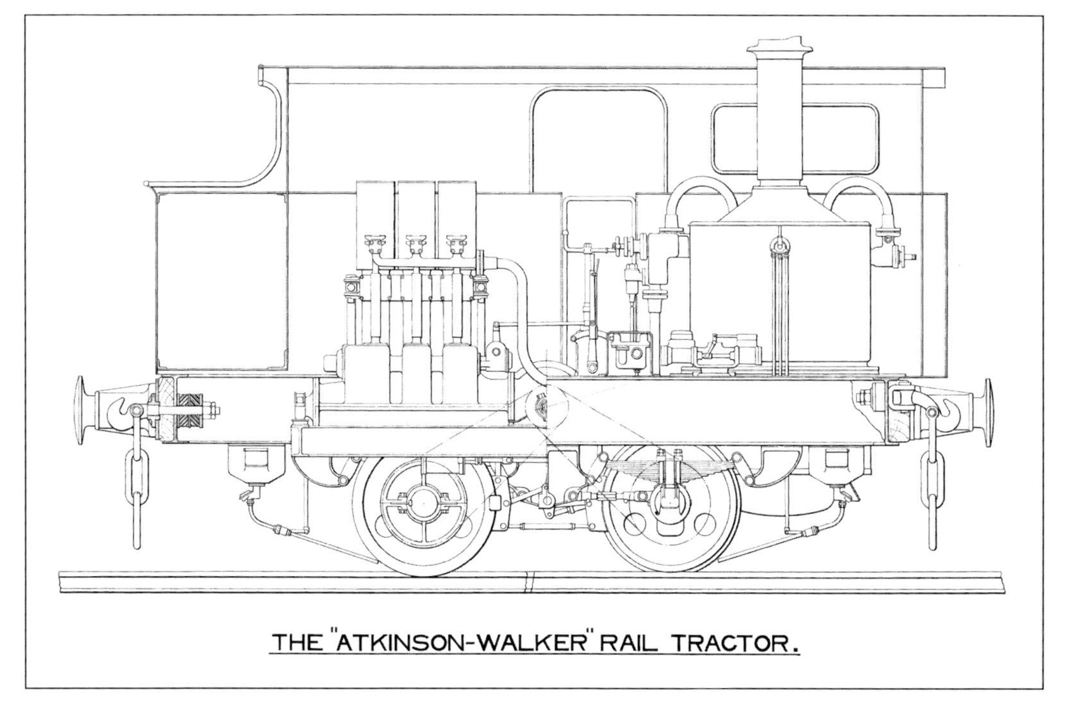

Figure 41: A general diagrammatic drawing of the four-wheel Atkinson-Walker three-cylinder steam rail tractor locomotive to give the reader a general idea of the layout of these vertical boiler locomotives.
Author's collection original acquired from an international auction site.

Locomotive details and dimensions for class B type AtW

Length over buffers	21 ft
Overall height	11 ft 6 ins
Overall width	8 ft
Boiler heating surface	125 sq. ft
Working boiler pressure	280 lb per sq. in.
Total grate area	7 sq. ft.
3 Vertical cylinders	7 in. bore x 10 in. stroke
Coal capacity	6 cwt.
Water capacity	1000 gallons
Weight working	35 tons
Wheel diameter	3 ft
Wheel base	5 ft 6 ins
Power output at 150 rpm	5700 lb tractive effort

Included above is a typical drawing of the four-wheel Class B, as no drawing of the Class A or Class D, as used by Oxford & Shipton Cement, has yet been found

AtW No. 109, running as Shipton's No. 2, was a two-cylinder – 7in x 10in cylinders, standard gauge, four-wheel and having 3ft diameter wheels. It was transferred to Alpha Cement Ltd. when the works was sold in June 1934.Then it left Shipton by 16th of July 1937 to Jabez Barker & Son Ltd., who were dealers, New Brent Street, Hendon, Middlesex. It was recorded still for sale there on the 10th of May 1940. However, it was then for sale at Goerge Cohen Sons & Co. Ltd. on 27th of December 1940. By 1941 (at the earliest) it was seen at Francis T. Wright Ltd., Finedon Railway Wagon Works, Northamptonshire and was scrapped there in October 1950.

When new, the locomotive was equipped with low-height buffers for hauling the original design of V-tipper wagons used to haul stone from the quarry face into the Shipton works crushing plant.

Plate 85: The Atkinson-Walker locomotive, works number 109 seen here with a sales advertising livery lined out and lettered for the Oxford & Shipton Ltd. company – all cleverly airbrushed on an original works photograph. Below is the works plate from the above locomotive.
Courtesy the Industrial Locomotive Society.

Plate 86: Oxford & Shipton Cement Limited Locomotive No. 2 (AtW 109/1928) seen here at F.T. Wright Ltd., Finedon Railway Wagon Works, Northamptonshire in July 1950.
Industrial Railway Society, photograph by G. Alliez.

AtW No. 110, running as Shipton's No. 1, was a one-off four-cylinder – 7in x 10in, standard gauge six-wheel and having 3ft diameter wheels. It was transferred to Alpha Cement Ltd. when the works was sold in June 1934. Then it left Shipton and was transferred to another Alpha Cement Ltd. plant at Metropolitan cement works, Thurrock, about 1936-37 and named GOLIATH at sometime during its life. The AGM states that this was moved for heavy repairs but that never seemed to have happened and it was stored in the open next to the quarry face and eventually was scrapped in July 1952.

Plate 87: The only class 'D' six-wheel locomotive ever built by Atkinson-Walker and sold new to the Oxford & Shipton Cement Limited in 1928, seen here in an official works airbrushed sales photograph. The livery is almost certainly not the livery that Oxford & Shipton Cement Ltd. chose and would have been just applied for advertising purposes. The locomotive, (works number 110 of 1928 – four cylinder) went on to Alpha Cement's Metropolitan works, West Thurrock, Essex in 1936-1937.
Courtesy Industrial Railway Society.

Plate 88: The Atkinson-Walker vertical boiler six-wheel locomotive GOLIATH seemingly dumped at quarry face in Metropolitan works, for supposedly repair in late 1937. On 26th of November 1947 (when this photograph was taken) it was still with its yellow painted name that was applied when at the Shipton-on-Cherwell plant.
Courtesy Industrial Railway Society.

Plate 88a: The last view of GOLIATH part dismantled laying in the Metropolitan works.
Courtesy Industrial Railway Society.

A LOOK NOW INTO THE CONVENTIONAL STEAM LOCOMOTIVES OF THE SHIPTON-ON-CHERWELL CEMENT WORKS

Plate 89: A rare panoramic view of three of the Shipton cement works steam locomotive fleet with (from left to right) No. 3 (AB 2041/1937), CFS No. 6 (RSH 7742/1952) and WESTMINSTER No. 5 (P 1378/1914) all lined up in an area that was obviously the coaling point situated near to the factory buildings. Note that they are all facing the same way, as no turntable existed within the works complex. At one period, however, there was a triangle within the quarry railway track that could have been used for turning the locomotives, although it is not known if it was ever used.

Courtesy John Scholes collection, photographer W. Potter 7/10/1967.

LOCOMOTIVE ALBION, LATER No. 1 – 0-4-0ST

ANDREW BARCLAY WORKS NUMBER 776 BUILT 1896

Plate 90: Locomotive ALBION (AB 776/1896 – later No. 1). The first of the conventional steam locomotives purchased by the company, standing in front of the quarry face which can be seen behind.
Courtesy Industrial Locomotive Society collection; Photographer Derek Stoyel 31/07/1933.

Plate 91: ALBION No. 1 (AB 776/1896) locomotive standing at the weather proofed water column, with the Shipton quarry face in the background. Note the wooden ladder (propped up against the water column), used to access the locomotive water fillers on the locomotive tank tops.
Courtesy Industrial Railway Society photographer B.D. Stoyel 1/1961.

Technical details of the locomotive:

Builder	Andrew Barclay, Kilmarnock
Works Number	776 built 1896
Class – Shunter	0-4-0ST
Cylinders	14 ins x 22 ins
Driving Wheels Diameter	3 ft 0 ins
Weight	26 tons
Boiler Pressure	160 psi
Tractive Effort	16,290 lb

This Andrew Barclay locomotive was built at the company's Kilmarnock works in 1896 for stock. It was eventually sold as new in November 1899 to the Mellingriffith Co. Ltd. Tin Plate works at Whitchurch, Glamorgan and named FIREFLY. It was sold in 1928 to P. Baker Ltd. (dealers) Cardiff and renamed ALBION.

Bakers then rebuilt this locomotive in 1928 and resold it to Oxford & Shipton Cement Ltd., in 1931 where it was still carrying the name ALBION. This was the first conventional steam locomotive that the Oxford & Shipton Cement Limited purchased.

After the Oxford & Shipton Ltd. became part of Associated Portland Cement Manufacturers, the locomotive was transferred in 1952 to one of their other plants at Houghton Regis, Dunstable, Bedfordshire.

Whilst at Dunstable, in the 1960s, it was renamed PUNCH HULL, to honour one of the plant's drivers and was later replaced by a modern diesel locomotive.

Made redundant in March 1967, it was presented for preservation to the London Railway Preservation Society and seen at BR Dunstable Town Station in storage. It was then moved on to Skimpot Lane Depot, Luton in October 1967 and then again on to Buckinghamshire Railway Centre, Quainton Road in April 1969. Under new ownership, it was relocated in July 1987 to the Rutland Railway Museum, Cottesmore and restoration started to bring it back into use with the original name of FIREFLY. Several changes again in ownership saw it arrive at the Pontypool & Blaenavon Railway in February 2016 but, by March 2017, it appears at the Northampton & Lamport Railway, Chapel Brampton.

Figure 42: The next three official plans from the Glasgow University Archives are all that was left of the AB776 works drawings and have been carefully reconstructed to be included in the book, from the poor copies of the remains.

Reproduced by permission of the National Records of Scotland and University of Glasgow Archives and Special collections. Andrew Barclay Sons & Co. Ltd. collection, GB248 RHP54604.

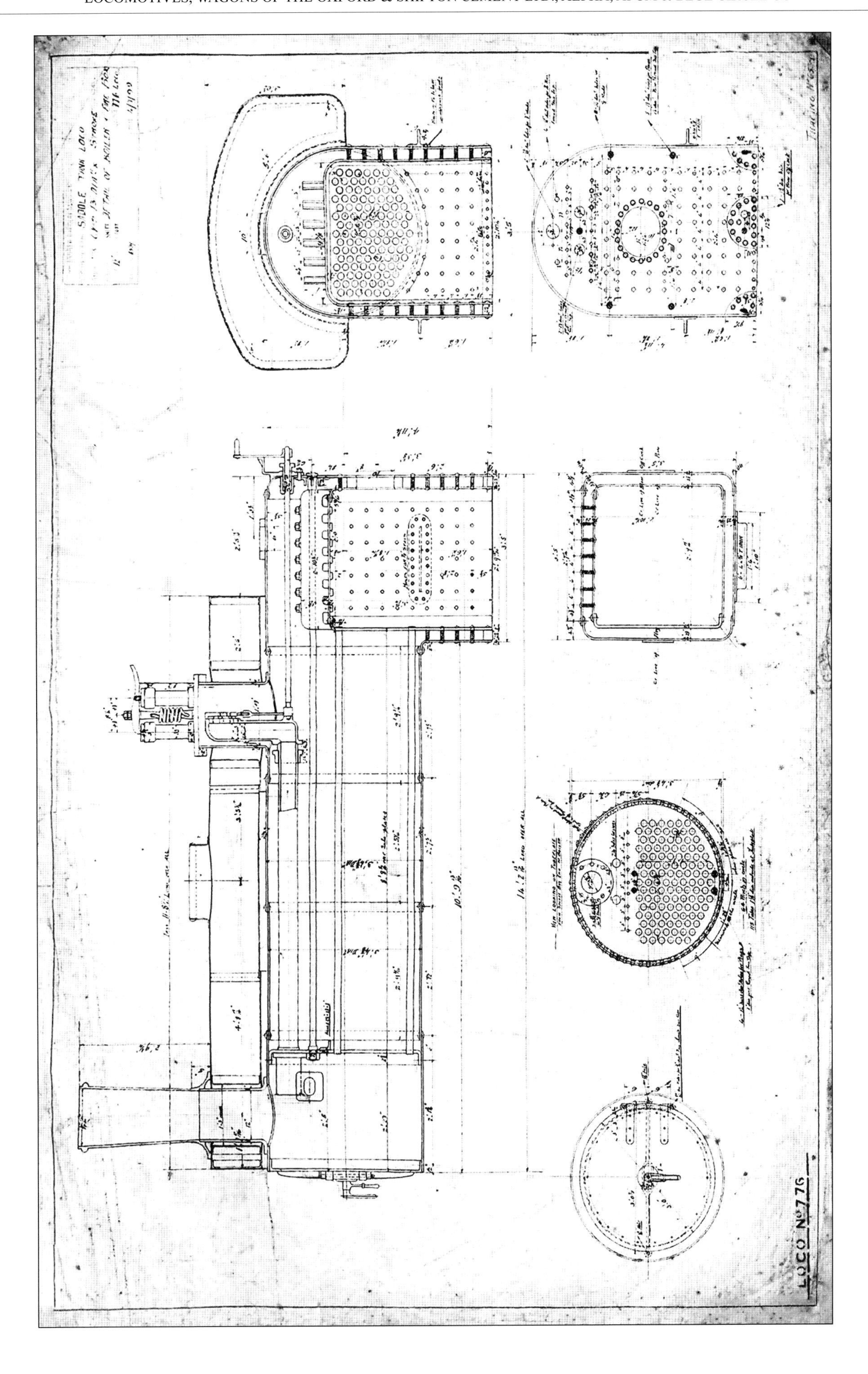
LOCO No 776

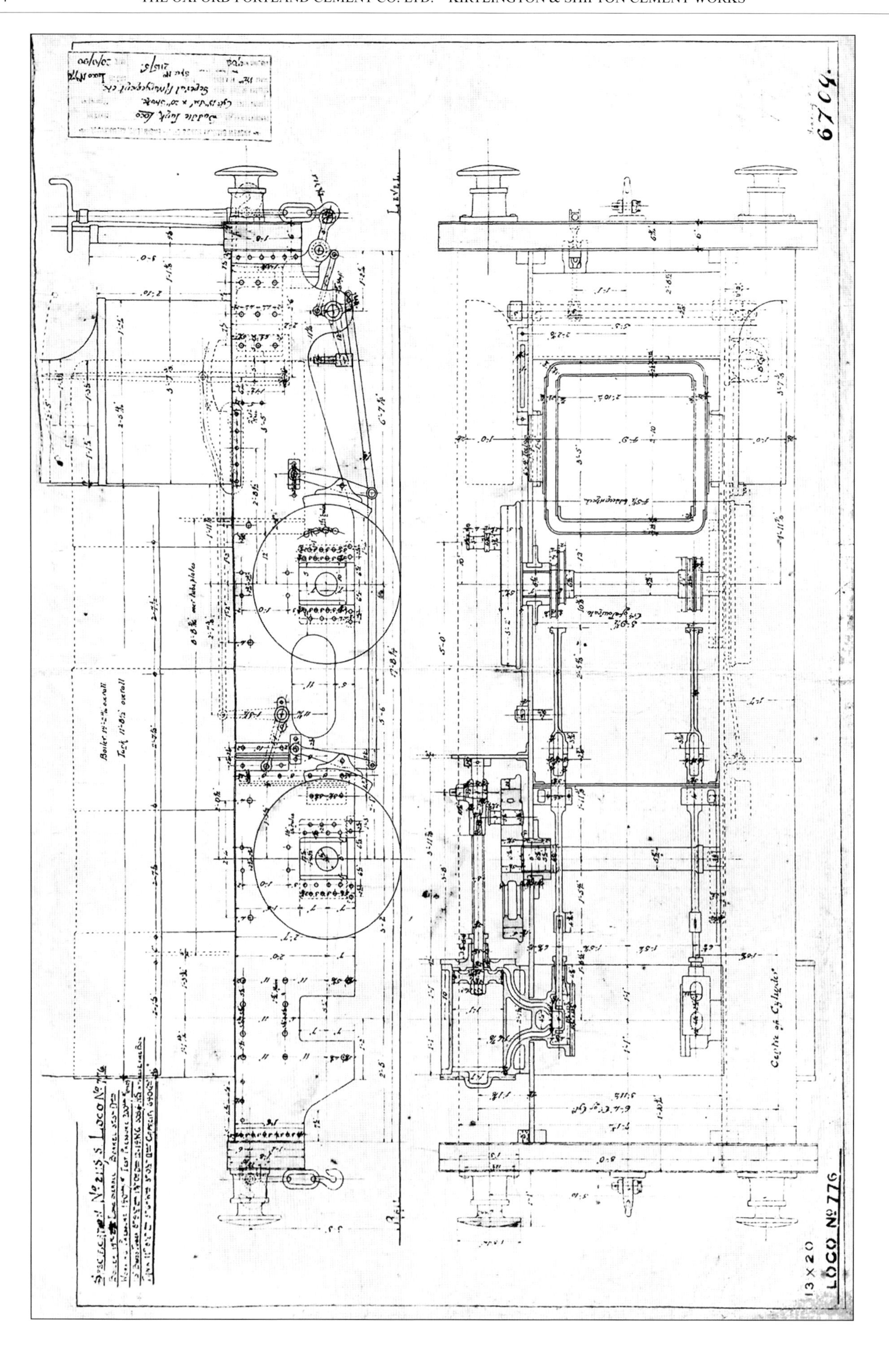
13 x 20
LOCO No 776
Centre of Cylinders

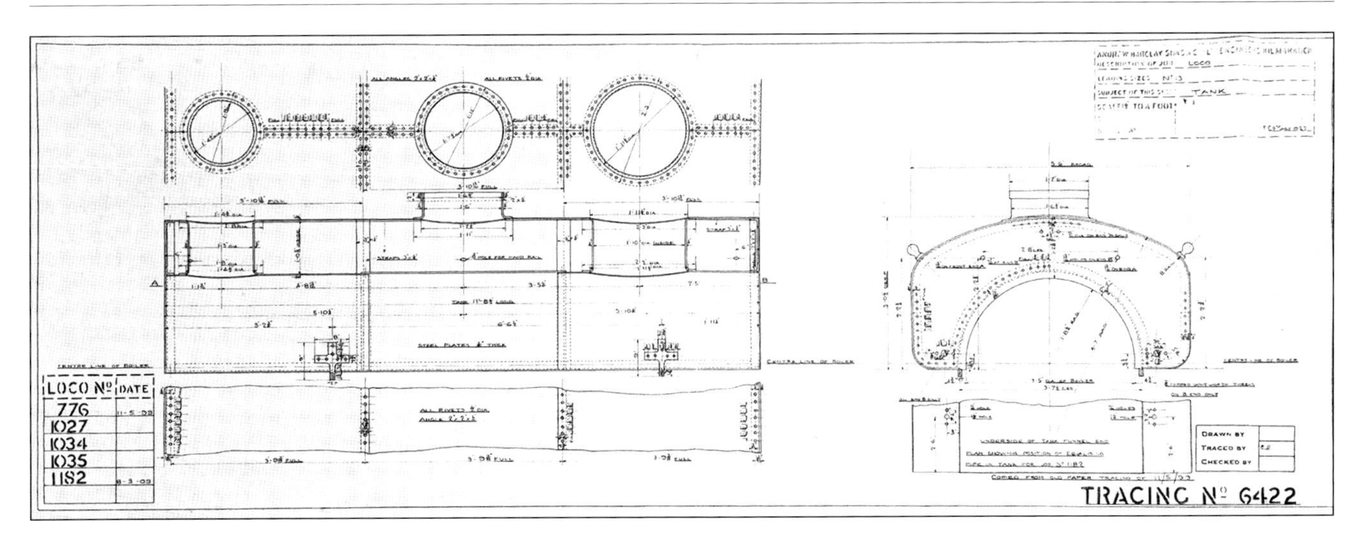

Figure 42a: A simple outline drawing of locomotive AB 776/1896.
Courtesy Industrial Railway Society.

Plate 92: A fine view AB 776/1896, one of Shipton's first locomotives, now with No. 1 painted on the side, with a couple of the later more box-shaped stone wagons in tow and supporting different design of wooden block buffers to those that were fitted to the later engines.
Industrial Locomotive Society; Photographer Alan E. Tyler 05/08/1954.

Plate 93: Locomotive AB 776/1896, seen here with its new name PUNCH HULL on the cab side, standing outside the APCM Ltd., Dunstable works locomotive shed in Bedfordshire.
Courtesy Industrial Locomotive Society collection; Photographer J.F. Clay.

Plate 94: A good front view of AB 776/1896 standing outside the Dunstable cement works shed just before it passed into preservation and presented to the London Railway Preservation Society in March 1967. Note all the wooden buffers have now been removed. TOM PARRY (AB 2015/1935) Andrew Barclay locomotive stands inside the shed.
Courtesy Industrial Railway Society, photographer Kevin Lane April 1965.

Plate 95: A good view of the back of the locomotive AB 776/1896, which is seldom captured on film. The loaded coal wagon on the left appears to have very fine coal lumps on board possibly for the works boilers.
Courtesy Industrial Locomotive Society collection; Photographer G.E. Hughes 19 July 1954.

Plate 96: A side view of AB 776/1896 showing an attitude that looks like the rear driving wheel springs are faulty or that the water tanks are empty? Seen whilst standing at APCM Ltd., Dunstable.
Courtesy Industrial Locomotive Society collection; Photographer Derek Stoyel 19/6/1954.

Plate 97: A last view at Dunstable of AB 776/1896 PUNCH HULL outside the APCM Ltd., Dunstable, Bedfordshire locomotive shed in 1964 taking on water. Now fitted with conventional buffers.
Courtesy Industrial Locomotive Society collection; Photographer J.M. Hutchings 04/10/1964.

Plate 98: AB 776/1896 standing at Skimpot Lane Depot, Dunstable of the London Railway Preservation Society, with an ex-LNWR coach standing behind it, also awaiting its turn for restoration.
Courtesy Industrial Locomotive Society collection; Photographer L.H. Somerfield c1968.

Plate 99: Long past being entering preservation; AB 776/1896 stands now in a sorry state, still awaiting restoration by the owners. Note that the wooden buffer beams have been removed completely (or rotted away), leaving the front and back steel facing plates. *Courtesy John Wolley collection.*

LOCOMOTIVE No. 2 NAMED ALBION II, LATER No. 2 W.G. BAGNALL WORKS NUMBER 2178 BUILT 1921

This locomotive was the last batch (WB 2171-8/1921) ordered by the India Office Development Department for the Salsette – Trombay Railway, Bombay. Numbered STR No. 8, it and its sisters cost £2,162 10s 0d each. As was normal for tropical countries, it had a double skinned roof *(see Plate 100)*, an unusual feature when it was returned to the UK.

The locomotive specification, as quoted in the Bagnall records are:

Type 0-4-0 Saddle Tank, outside cylinders, inside frames, double roofed cab.
Cylinders 12” x 18”
Driving wheels 3 ft x 0½ ins dia. wheels.
Wheel base 5 ft x 3 ins.
Heating Area HS tubes 448sq ft, fire box 46 sq ft, grate area 7.5 sq ft.
Boiler Pressure160 lbs per sq ins.
Water tank capacity550 gallons – boiler capacity 16 cu. ft.
Tractive effort at 85% BP – 9,658 lbs.
Weight 21 tons loaded – 16.7 tons tare

Purchased by Thos. W. Ward, dealers, Greys, Essex c1929, it was brought back to the UK. It was then purchased by the Cement Industries Ltd. (later Alpha Cement Ltd.) at Rodmell, near Lewes, Sussex in 1933 and named ALBION. The locomotive then appears at Shipton-on-Cherwell cement works c1937 and given the name ALBION II (later No. 2). The official order book states it was noted still working there on 8th of September 1963. The Yorkshire Dales Railway acquired it in September 1971 for preservation but by October 1972 it was offered for sale for £150, but since it has been scrapped and cut-up.

The general orders for spares from the Bagnall works for this locomotive are listed as follows:

Order No. 9770 dated 08/04/1931	spares 2178	T.W. Ward, Titan Works, Greys Quarry Sidings.
Order No. 3501 dated 11/06/1931	spares	T.W. Ward, Titan Works, Greys Quarry Sidings.
Order No. 5864 dated 23/06/1933	spares	Cement Industries Ltd., Rodmell Works nr Lewis.
Order No. 4394 dated 23/02/1937	spares	Alpha Cement Co., Bletchington, Oxfordshire.
Order No. 3590 dated 25/11/1940	spares	Alpha Cement Co., Bletchington, Oxfordshire.

We are lucky to have the Bagnall Works archives housed at the Staffordshire Record Office so that we can include more details on this locomotive than on the others in the Shipton fleet. After the initial design stage of the engine, the company would need to draw every part of the locomotive in minute detail for the workshops to be able to manufacture all the components.

Plate 100: With the protective weather curtains closed, Bagnall No. 2178/1921 stands outside the single road Shipton works locomotive shed.
Courtesy Industrial Railway Society, photographer A.R. Etherington.

238

4. WHEELS COUPLED. 5'-3" WHEELBASE. 4'-8½" GAUGE

10007	102	PRESSURE GAUGE, WHISTLE & CYLINDER COCKS		1	8-9-21
10010	"	OIL CUPS & MUD PLUGS		1	" "
9988	"	1" WHEEL VALVE		1	" "
10946	113	BRAKE SHAFT & BRACKETS	x	15	9-9-21
10956	"	CHIMNEY	x	15	" "
10955	"	EXHAUST PIPES	x	15	" "
10960.	"	Trailing Sandbox & Details	x	15.	12/9/21.
10959	"	Frame Outline	x	"	"
10022.	73.	Regulator Stuffing Box, Rod & Handle.		2.	" " "
~~10960~~	~~73~~	~~TRAILING SANDBOX & DETAILS~~		~~15~~	~~14/9/21~~
10957.	113.	Steam Pipes, Flanges etc.	x	15.	" " "
10964	"	Smokebox Door Details	x	"	"
10966	"	Regulator Valve	x	"	"
10967	"	Regulator details	x	"	"
10965.	113.	Stay plates & Slidebar Suppt.	x	15.	16/9/21
10963	"	ANGLE IRON LIST & FOOTSTEPS		15	19/9/21
10952	"	REVERSING LEVER & SECTOR	x	"	"
10977	"	ASHPAN & DETAILS		"	27/9/21
10962	"	LEADING BUFFER BEAM & STAYS		"	" "
10976	"	CAB. TANK & BUNKERS.		"	
10978	"	FIREDOOR & DETAILS		13	
10974	"	DRAWGEAR DETAILS		15	
10968	"	LEADING SANDBOX & DETAILS		"	
10979	"	CAB & TANK DETAILS		"	
10972	"	SMOKEBOX ARRANGEMENT		"	
10981	"	CYLINDER COCK GEAR		"	30-9-21
10986	"	CAB DETAILS.		"	
10987	"	FRAME ARRANGEMENT		"	
10980	"	REAR BUFFER BEAM & STAYS		"	
10991	"	SMOKEBOX, WATER COCK, & JACK CLIP DETAILS		"	
10994	"	LAMP BRACKETS.		"	
11014	"	PIPE ARRANGEMENT		"	12-10-21
11002	"	WATER & PRESSURE GAUGE, INJECTOR & WHISTLE		"	
10990	"	NUMBER PLATE		"	15-10-21
6849	73	TOOLS & OIL CANS ETC		2	8-1

237

ENGINE No 2171-8 CYLS 12" x 18"

10921	113	STEEL PLATES	x	15	16-8
10916	113	COPPER PLATES	x	15	18-8
5646	75	TEE PIECE		1	"
8131	88	CLACKBOX & SEATING		~~15~~	"
5769	76	CYLINDER & SLIDE VALVE		15	"
5761	"	CYLINDER COVERS, PISTON & GLANDS		13	"
5711	"	CROSSHEAD & PISTON ROD		15	"
5704	"	MOTION DETAILS		15	"
3048	"	ECCENTRIC DIAGRAM		15	"
3197	"	ECCENTRICS		15	"
10982	113	VALVE MOTION ARRANGEMENT		15	"
5760	76	AXLEBOX & GUIDES		13	"
6004	73	TANK FILLER		1	"
10930	113	WHEELS, AXLES & CRANKPINS	x	15	24-8
10934	113	SPRINGS & DETAILS	x	15	30-8
10933	113	CONNECTING & COUPLING RODS	x	15	31-8
11083	113	SAFETY VALVE		1	5-9
10949	113	MOULDINGS & CASINGS	x	15	"
8302	88	STEAM STAND		13	"
10000	102	STEAM BRAKE VALVE.		1	"
5752	75	BRAKE BLOCKS		1	"
10947	102	FOUNDATION RING		1	"
10925	113	BOILER & FIREBOX	x	15	7-9
10950	"	REVERSING SHAFT & BKTS	x	15	"
5709	76	BRAKE COLUMN & SCREW		11	"
10953	113	VALVE RODS & SLIDEBARS	x	15	"
5710	76	STEAM BRAKE CYLINDER		15	"
10951	113	EXPANSION BRACKET & BOILER STAYS	x	15	"
10948	113	BRAKE GEAR DETAILS	x	15	"
10187	102	BLOWER VALVE & ROD		1	"
9814	"	DRIVERS SHELF		1	"
9976	"	BLOW OFF COCK		1	"
5680	75	CAB WINDOWS		1	"
11018	113	PIPE ARRANGEMENT DETAILS		15	20/10

Figure 43: The two pages from the Bagnall drawing register showing all the individual drawings needed to build locomotive works No. 2178/1921. Reproduced courtesy of the Staffordshire Record Office.

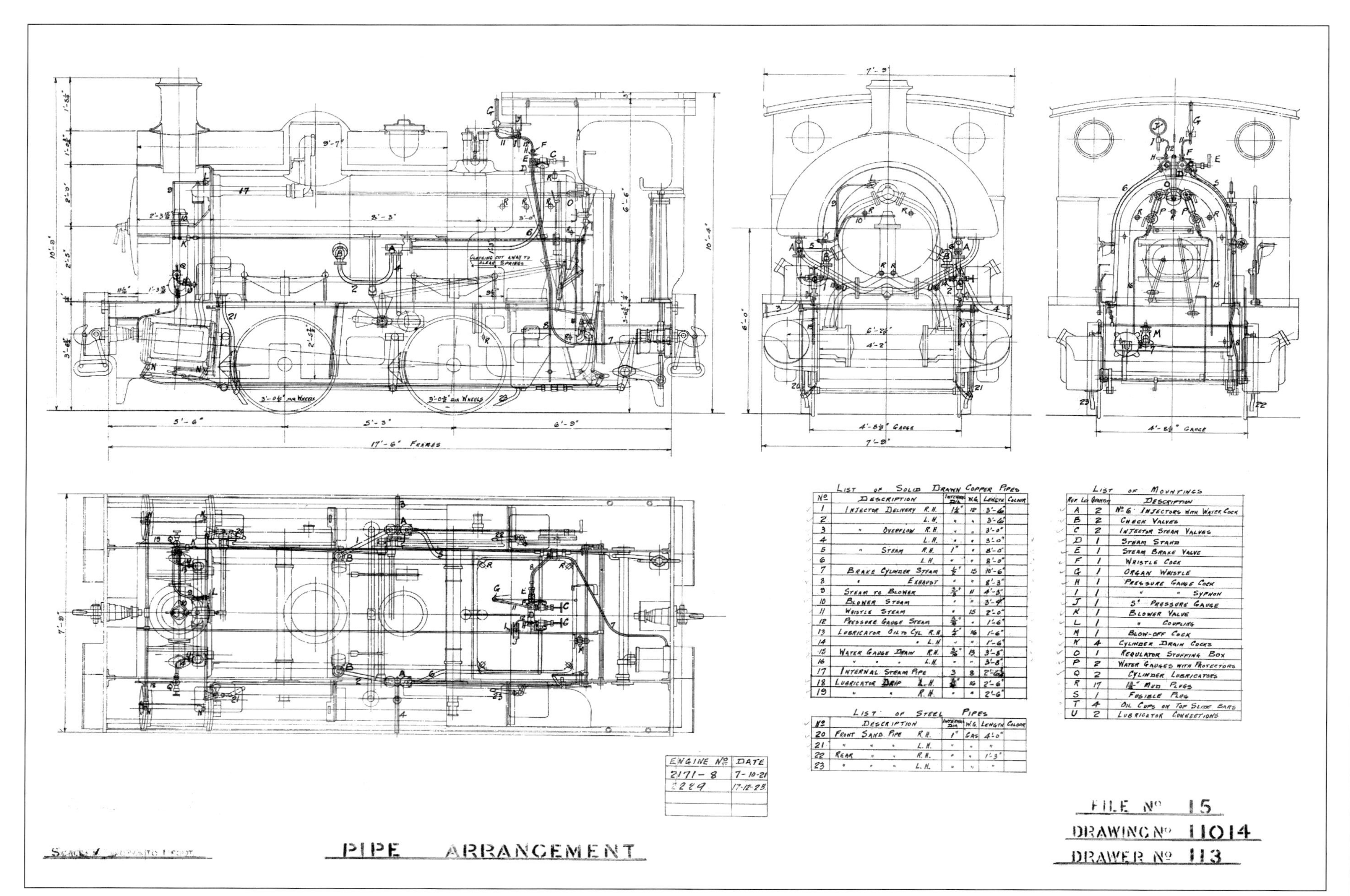

Figure 44: The next four drawings are from the Bagnall Company construction drawings used on the batch of locomotives including WB 2178/1921. Reproduced courtesy of the Staffordshire Record Office.

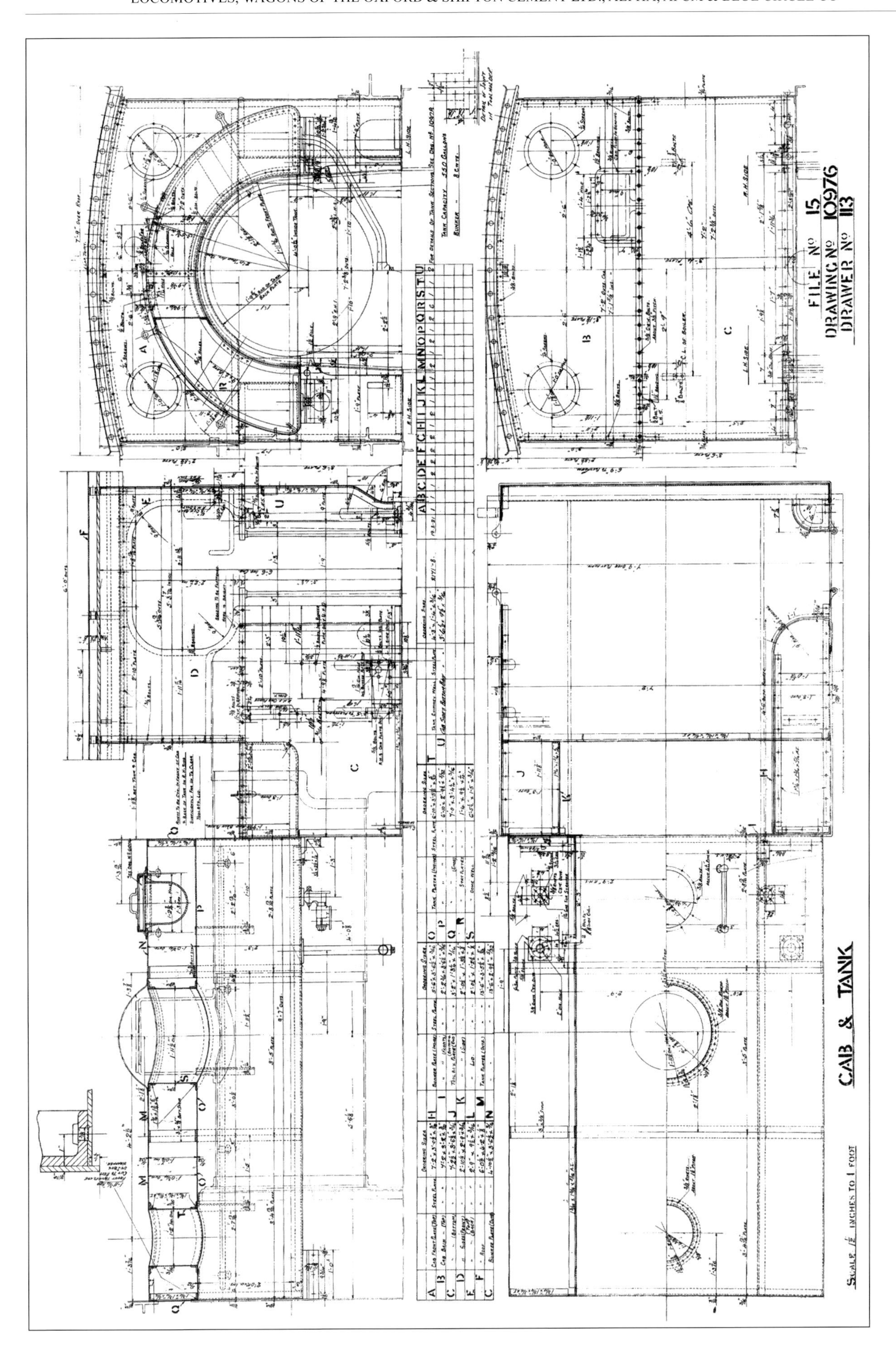
FILE No 15
DRAWING No 10976
DRAWER No 113
TANK CAPACITY 550 GALLONS
BUNKER - 8 CWTS
CAB & TANK
SCALE 1/12 INCHES TO 1 FOOT

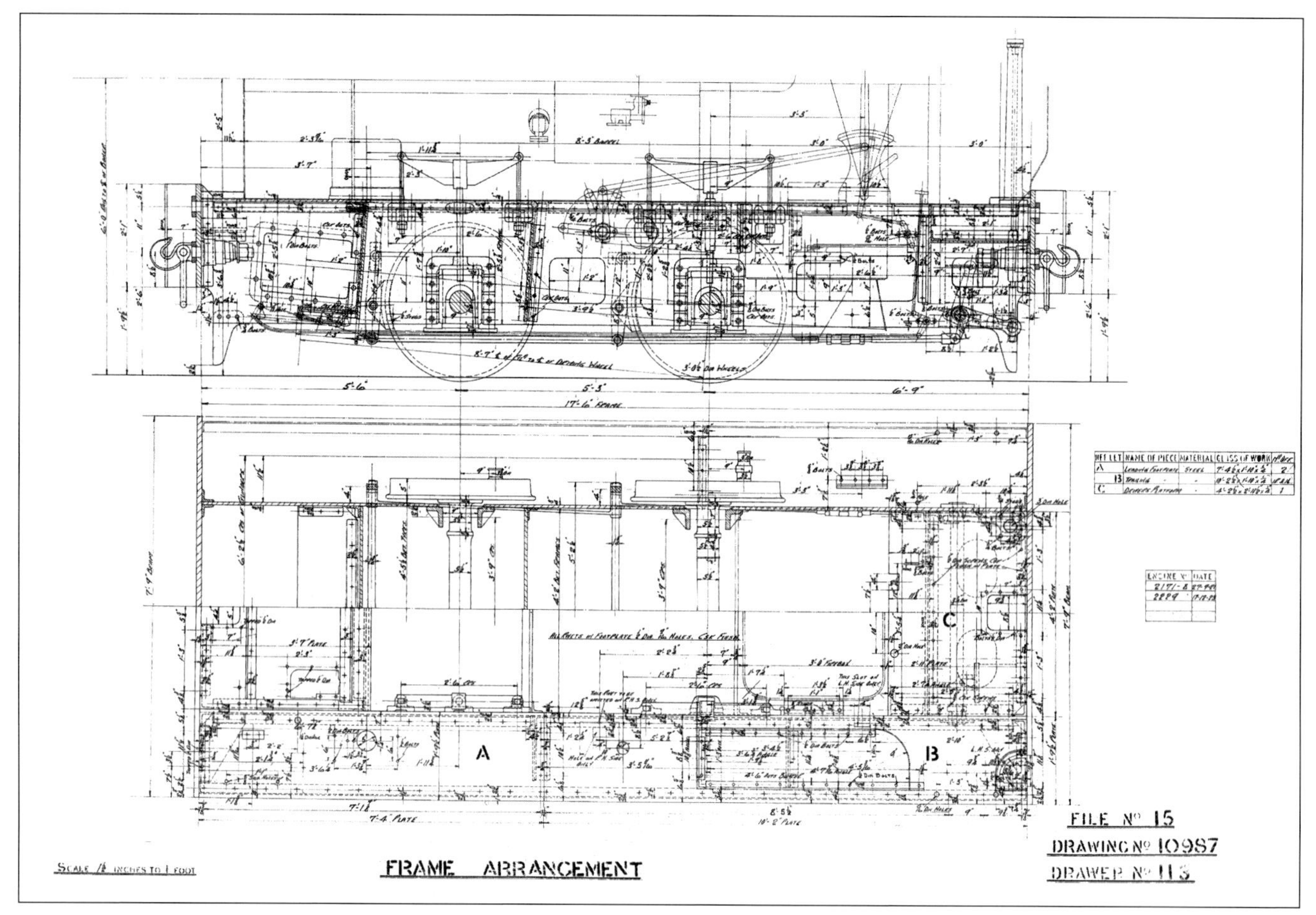
A
B
C
FILE Nº 15
DRAWING Nº 10987
DRAWER Nº 113
FRAME ARRANGEMENT

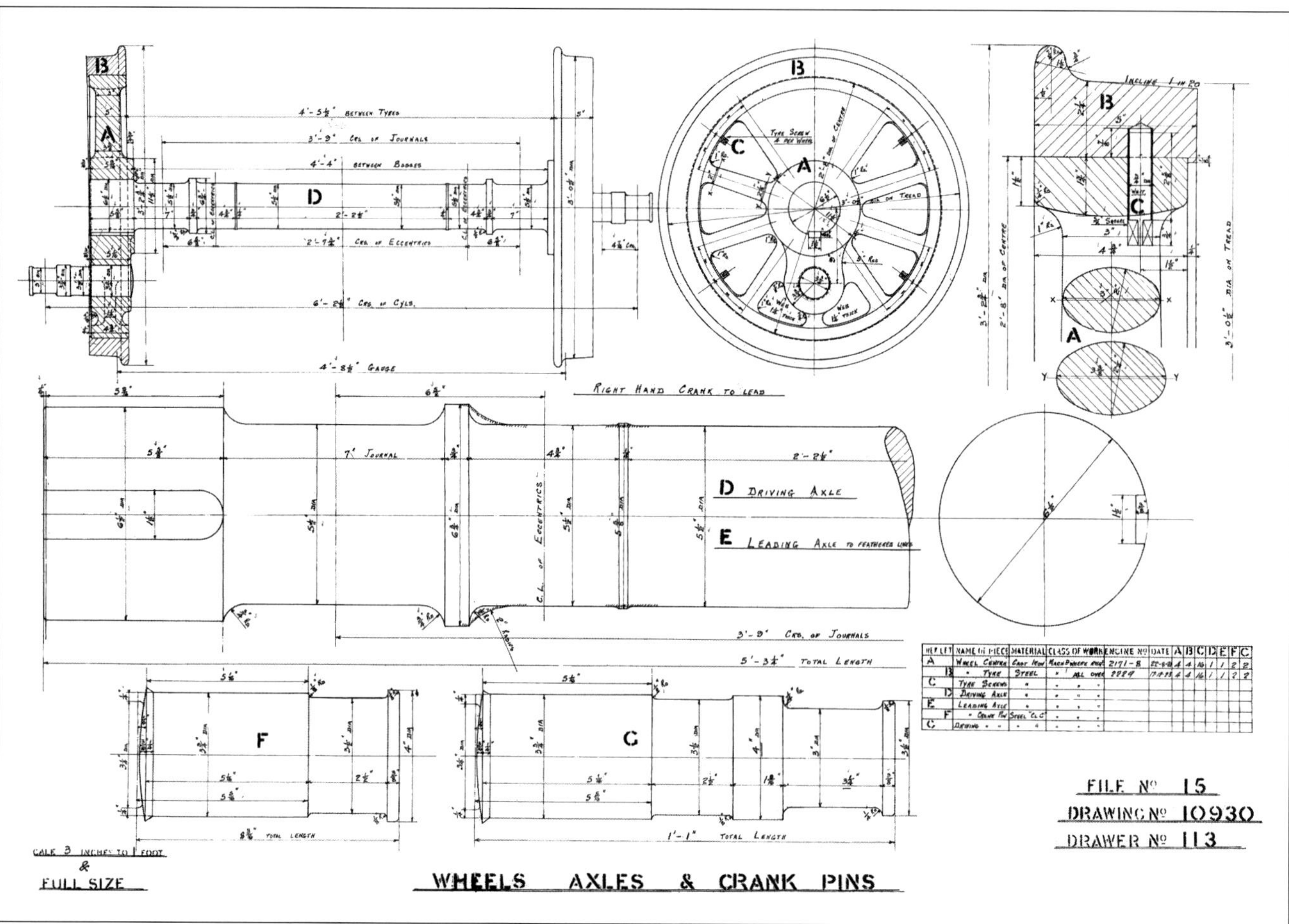
A
B
C
D
RIGHT HAND CRANK TO LEAD
D DRIVING AXLE
E LEADING AXLE
F
C
FILE Nº 15
DRAWING Nº 10930
DRAWER Nº 113
FULL SIZE
WHEELS AXLES & CRANK PINS

Plate 101: WB 2178/1921 simmering in steam, outside the locomotive shed at Shipton on 30/9/1967. The wire fence in the foreground marks the divide between BR and APCM property.
Courtesy Industrial Railway Society; photographer J.P. Mullett.

Plate 102: The second locomotive, No. 2 (WB 2178/1921), seen at the Shipton plant works sidings with an Eastern region van probably loaded with bagged cement.
Courtesy Industrial Locomotive Society collection; Photographer G. Alliez 19/08/1953.

Plate 103: A rare rear view (looking south), of WB 2178 standing near the junction between GWR/BR and APCM property as marked by the wire fence – the BR (ex GWR) mainlines are in the far distance. Note the extension to the cab to form an external coal bunker that was added whilst at Shipton.

Courtesy Geoff Broadhurst on a private visit on 8th June 1967.

Plate 104: Maintenance taking place on Bagnall No. 2 whilst standing on one of the internal tracks. Note the massive works conveyor in the background.
Courtesy Laurence Waters collection; Great Western Trust collection, Didcot.

Plate 105: Here locomotive No. 2 (WB 2178/1921) shunting during 1971 before being sold in September into preservation at the Yorkshire Dales Railway at Embsay until finally scrapped.
Courtesy Laurence Waters collection; Great Western Trust collection, Didcot.

Plate 106: A good view of the front of locomotive WB 2178/1921, showing the huge buffer blocks, fitted to accommodate all types of wagons buffer heights.
Courtesy Industrial Locomotive Society collection; Photographer H.A. Gamble 20/04/1965.

Plate 107: A fine colour view of WB 2178/1921, as Shipton works No. 2, outside the single road locomotive shed, on the private visit of 8th June 1967. Again, the BR mainlines and Shipton exchange sidings are visible on the right.
Courtesy Geoff Broadhurst.

Plate 108: A sorry looking WB 2178/1921 now in preservation at the Yorkshire Dales Railway Society, Embsay Depot, West Riding of Yorkshire. Study of this view and the original drawings shows that the extended coal bunker was a later addition, whilst at Shipton. Wonder what the B1 stands for?

Courtesy Industrial Locomotive Society collection; Photographer John Meridith 01/07/1972.

LOCOMOTIVE III, later No. 3 – 0-6-0ST

Andrew Barclay works number 2041 built 1937

This locomotive was delivered NEW to Shipton-on-Cherwell cement works in 1937. It worked for the Alpha Cement Co. Ltd. and all the successor companies at Shipton until 1971, when it was scrapped and cut up on site by J. Friswell & Son, Banbury in September 1971 after 37 years of hard work!

Plate 109: The official works photograph of locomotive Andrew Barclay works number 2041/1937. This is a poor reproduction from the damaged glass plate that has not weathered the time well, but worth including as it bears the number III on its saddle tank.

Reproduced by Permission of the National Records of Scotland and University of Glasgow Archives and Special Collections, Andrew Barclay Sons & Co. Ltd. collection, GD 329/6/1/2181.

Plate 110: No. 3 – AB 2041/1937 and crew! Note the wooden ladder hooked over the saddle-tank hand rail which could be moved along and/or across to the other side of the locomotive. This seems very practical and sensible arrangement and is quite a rarity in the UK. The livery is also different as the whole locomotive is painted this apple green and no sign of red except on the buffer beam.

Courtesy John Scholes collection, photographer A.N. Glover 4/11/1961.

Plate 111: A view in the Shipton quarry with locomotive No. 3 -AB 2041/1937at the head of a train of wagons being loaded with stone by the R&R 5160 electric excavator and shovel during quarry operations.
Courtesy John Scholes collection, photographer A.N. Glover 4/11/1961.

Plate 112: Another fine view of No. 3 running along one of the quarry lines propelling the solid wheeled V-shaped wagons. This view shows that these quarry lines were regularly altered as the quarry face receded, as the row of sleepers show in the foreground.

Courtesy Industrial Railway Society.

Plate 113: Good example of how the large buffer plates work is shown in this view of Andrew Barclay locomotive 2041/1937 with an internal steel box shaped wagon in tow.
Courtesy Industrial Locomotive Society collection; Photographer G. Alliez 17/05/1958.

Plate 114: A fine colour view of two of the Shipton steam fleet in very poor condition – AB 2041/1937 and WESTMINSTER on a spur in the quarry internal rail system. Note WESTMINSTER seems to be out of use with its chimney covered up and AB 2041/1937 has had a nasty collision at the front with the side of the running plate buckled.
Courtesy Gordon Edgar.

Plate 115: Having just brought a full load up from the quarry face, an almost rusty AB 2041/1937 stands waiting for the next move.
Courtesy Geoff Broadhurst 8/6/1967.

Plate 116: Locomotive AB 2041/1937 standing alongside the braked internal wagons with the shunter just either applying or releasing the empty wagons brakes. Note how near the quarry face had become towards the Bunkers Hill houses in the distant, up on the Bletchington road. The enormous wooden buffers show up well in this image and also the steel protection bar that had been fitted under the locomotive to push any fallen rocks out of the way, thus avoiding damage to the engine.
Courtesy Joe Turners photograph.

LOCOMOTIVE No. 4 – 0-6-0ST

HUNSLET ENGINE WORKS No. 1649 BUILT 1930

This locomotive carried the wrong works Plate 1643 when new and was delivered to the Crown Agents, for the Halfa Harbour works construction contract in Palestine as H.H.W.D. No.7. Next seen stored with Epstein Strykowski at Halfa between completion of harbour contract and then appeared at Wadi Surar in July 1941, Rafa on November 1942 and then back to Wadi Surar in February 1943. Thence to Egypt at the 169 Workshops (Suez) in February 1944 named DOREEN. On to Ataka-Adablya line in July 1944 before returning to the UK in February 1948, when it went for short time to the Melborne Military Railway, Derbyshire, before arriving at the Alpha Cement Company, Shipton-on-Cherwell in August 1948 and given the Alpha Cement Company running number 4. It was scrapped on site by James Friswell & Sons Ltd., Banbury, Oxfordshire in January 1967.

FOR 44460
SEE E.N. 1506.

LIST OF LOCOMOTIVE ENGINES BUILT AT THE HUNSLET ENGINE WORKS, LEEDS.

Order No.	Makers' No.	Description of Engine	Coupled Wheels	Gauge	Type	Tank or Tender	Drawing Book	Tried in Steam	Sent Away	Name or Number Plate	Purchasers	Destination
C 44450.	1648	14 x 20 out cyls.	6, 3'-4"	4'-8½	0-6-0	Saddle	10/43647		29/5/31.	MILLBROOK	John Mowlem & Co. Ltd	Southampton Docks.
44330.	1649	14 x 20. ins cyls	6, 3'-4"	4'-8½	"	"	9/175[illegible]180		28/1/30.	No 4 loco. H.H.W.D. No. 7	Alpha Cement Co. Crown Agents. Indent 17/1 Palestine - Haifa. Harbour Works	[illegible]
49340. "	1650	" " " "	"	"	"	"	"		28/1/30.	R.N.P.F. No 1. H.H.W.D. No 8	Royal Naval Propellant Factory. (Admiralty) Per Freeman Fox & Partners.	Caerwent. Monmouth. " "
44340.	N 1651	16 x 22 " "	6, 3'-9"	"	"	"	10/48552		7/2/30.		The Peruvian Corporation Ltd. Southern Ry. of Peru.	Reqn No. 17A. Specn. V.419
"	1652	" " " "	"	"	"	"	"		19/2/30.		" " "	" " "
44800.	1653	15½ x 22. out cyls	6. 3'-6¾"	3'-3⅜	4-6-0	Tender.	10/15519		2/1/30.	S.P.P. 14 1929	Parana Plantations Ltd.	São Paulo. Parana Brazil
"	1654	" " " "	"	"	"	"	" "		12/2/30.	15.	" " "	" " "
C 44810.	N 1655	15 x 22 out cyls	6, 3'-7"	3'-3⅜	2-6-2	Side tanks.	10/84591		9/5/30.	K.U.R. 31	Crown Agents, Kenya Uganda Ry.	Reqn No. 6966
"	1656	" " " "	"	"	"	" "	" "		9/5/30.	" " " 32.	" " "	" " "
44330	1657	14 x 20 ins cyls	6, 3'-4"	4'-8½	0.6.0	Saddle	9/17558		26/2/30.	H.H.W.D. No 9	Crown Agents. Indent. 17/1 Palestine: Haifa Harbour Works.	
"	1658	" " " "	" "	"	"	"	"		26/2/30.	H.H.W.D. No 10	" "	" "
48890. 44440.	1659	14 x 20 ins cyls	6. 3'-4"	4'-8½	0.6.0	Saddle	9/17568		48890. 1938. 5/9/30.	H.H.W.D. No 11	The Director of Navy Contracts. 2.4.6. Bainbridge St. London. W.C.1. Crown Agents. Reqn No. [illegible] Palestine. Haifa Harbour [illegible]	
"	1660	" " " "	"	"	"	"	"		6/9/30	R.N.P.F. No 2. H.H.W.D. No 12	Royal Naval Propellant Factory. (Admiralty) Mr Freeman, Fox & Partners	Caerwent. [illegible]

Figure 45: Page extract from the Hunslet Engine book listing the locomotives built including works number 1649.
Courtesy Industrial Railway Society.

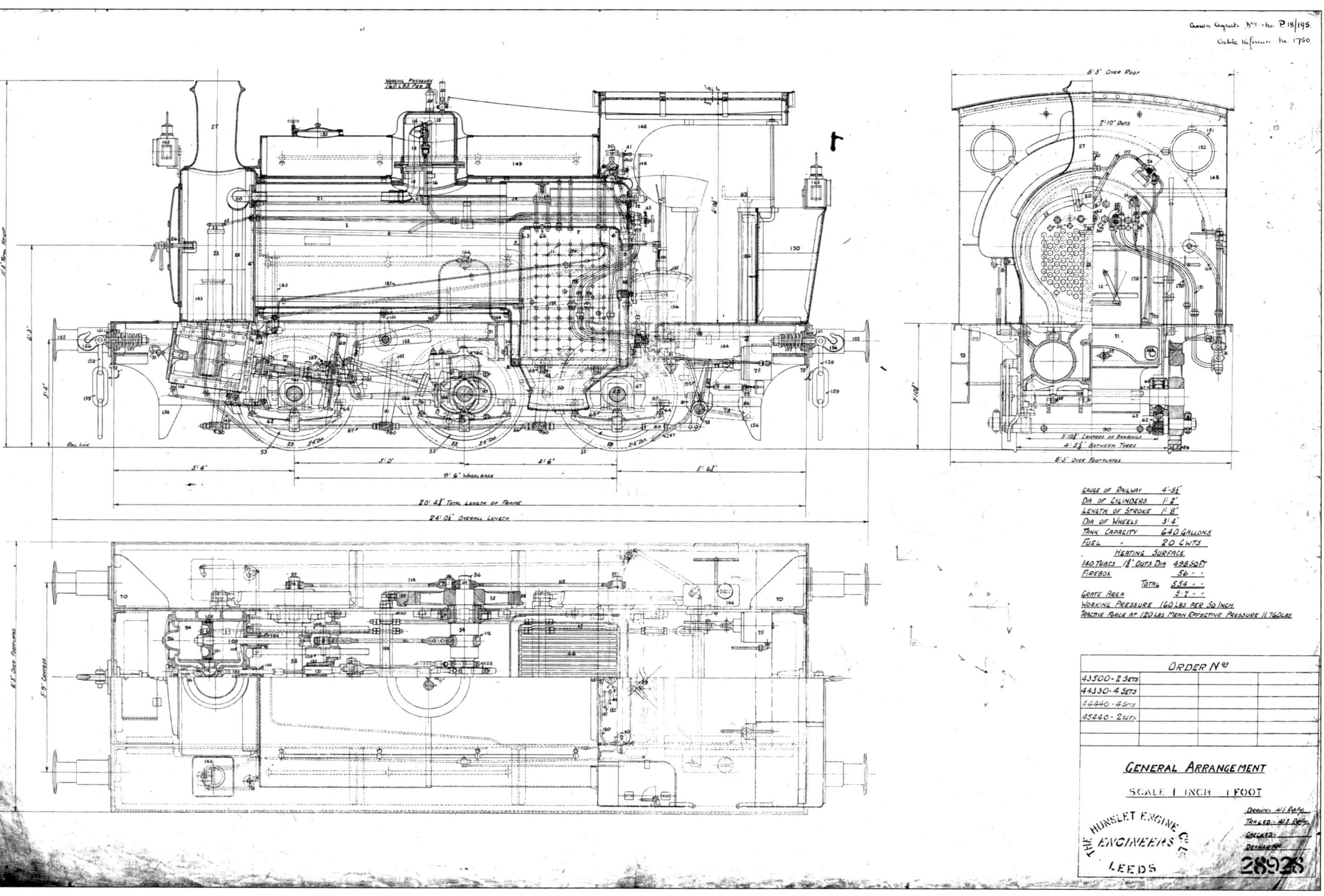

Figure 46: The General Arrangement drawing apllicable to Hunslet Engine locomotive works number 1649/1930.

Courtesy Industrial Railway Society.

44330

Estimate No. 29/14 29/25 29/135 Order No. 44330

Received	Issued	CUSTOMER'S No.	Date	Delivery	Marking Letters
4.12.29.	31.12.29.	Pal H.H.D.17/1 Ind.No. 17 dtd. 19.8.29.	23.8.29. 3.12.29.		
4.12.29.		Ditto		1st Jan 22nd 1930. 2nd Feb. 5th " 3rd & 4th Feb 25th 1930.	

Railway. Palestine Haifa Harbour Works, Ind. No.17 of 19.8.29. Dept. Pal. H.H.D.

Name and Address. The Crown Agents for the Colonies, 4, Millbank, London S.W.1.

Inspection Rendel Palmer & Tritton. Speon .. Teste

Engine No. 1649, 1650, 1657 and 1658 Class 0-6-0 type Saddle Tank Cylinders 14" x 20" inside.

Packing for shipment Carriage f.o.b. Manchester or equal.

S/S copies Inv. copies

Consign to:-

Order No. or file mark to be quoted on all communications.

Item No. Particulars of Order and Drawing Nos.

4 Locomotives with inside Cylinders and to the following specification:-

Gauge of Railway 4'-8½"

Cylinders 14" dia x 20" stroke

BOILER MACHINE ERECTING PLANNING DRAWING WORKS GENERAL COST TIME STORES.

Continuation Sheet No.1. Order No. 44330

Item No. Particulars of Order and Drawing Nos.

6 Coupled Wheels 3'-4" dia.

Length of Wheel Base 9'-6"

SPECIFICATION.

Boiler Plates - B.S.S.No.16. Acid Open Hearth Steel

Copper Firebox " " 11

Copper Boiler Tubes - B.S.S.No.13.

BOILER TESTS.

Hydraulic 240 lbs. per sq. in.

Steam 170 " " "

Working Pressure 160 lbs. per sq. in.

NOTE:- Boilers to have two coats of "Apexior" compound internally.

MOUNTINGS.

Two 1½" Ross Pop Safety Valves (on Dome Top)

Sliding type Regulator in Dome

Two Water Gauges 1/2" stock pattern with protectprs and Index Plates.

1 Steam Pressure Gauge & Cock

1 Deep tone Whistle on top of Dome.

2 No.7. Self-Acting Injectors placed under footplate.

2 Combination Valves on firebox back, for injectors.

1 Blow off Cock

1 Blower valve

1 set Mud Plugs and Doors

Sliding Firedoor

Continuatio Sheet No.2. Order No. 44330

Particulars of Order and Drawing Nos.

W.I. Firebars (stock section)

Steam Brake Valve (L & B Pattern)

Steam Stand

1 "Detroit" Single Feed S.F.Lubricator

BUFFERS AND DRAWGEAR.

Engine to be provided with a centre drawbar and spring on each beam. Each beam tp be provided with two buffers with 16" dia. heads.

WHEELS & AXLES.

All the wheel centres to be made of cast iron and fitted with steel tyres. Crankpins of steel class "A" casehardened.

AXLEBOXES & HORNBLOCKS.

Axleboxes to be of cast steel with phosphor bronze bearings.

No liners on faces.

Hornblocks of cast iron (single pattern not adjustable)

SPRINGS.

Laminated type and hung under the axleboxes.

BRAKE.

To act on all six wheels through cast iron blocks.

Steam and hand brake is to be fitted to be applicable either together or independently.

Hand Brake to be on left hand side.

Continuation Sheet No.3. Order No. 44330

Item No. Particulars of Order and Drawing Nos.

CYLINDERS.

To be 14" dia x 20" stroke, inside the frames.

Slide valves of cast iron

5 Drain Cocks to be fitted

Steel sheeting to be fitted on underside

No lagging required.

VALVE GEAR.

Stephenson's type inside the frames, all working parts of class "A" casehardened.

Crossheads of steel Class "C"

Slideblocks of cast iron

Piston Rods of steel Class "D" with steel nuts

Slidebars " " " "A" (4 bar type)

Eccentrics of cast iron

Eccentric straps of cast steel, lined with white metal.

COUPLING & CONNECTING RODS.

To be made of Steel Class "C" with bearings of phosphor bronze and big end of Connecting Rod lined with white metal.

Connecting Rods to have straps attached by through bolts, at both large and small ends.

Coupling Rod to be of the solid eye type with round bushes.

The Brasses for both ends of connecting rods are to be capable of adjustment by screw and wedge blocks.

Figure 47: This and the next page are the detailed particulars of the order for this locomotive. Unfortunately they had been defaced and crossed out across by the company, possibly after delivery but are now readable with all the details of each piece needed and also delivery instructions. Restored to their original state they make very interesting reading. *Courtesy Industrial Railway Society collection.*

Continuation Sheet No.4. Order No. 44330

Item No. Particulars of Order and Drawing Nos.

SADDLE TANK.

Capacity 640 gallons and provided with filling hole with hinged lid.

Rail Washing Cocks are to be fitted on Tank, with pipes to wheels and gear for operating cocks from the cab.

FUELBOX.

Placed at back end of engine. Capacity 45 cub.ft. - 20 cwt. coal.

CAB.

To have double roof, outer of steel, inner of wood (teak) to cover in the drivers footplate. Front plate to be fitted with 2 circular swivelling windows. The back of the cab to be open and the roof supported on pillars

SANDBOXES.

To be made of cast iron, 2 leading sandboxes placed on footplate, 2 trailing placed underneath footplate.

Front and hind gear to be worked by separate handles.

RAILGUARDS.

Bent from steel plate and placed two at each end.

BOILER CLOTHING.

To be steel sheets arranged on crinolines. No asbestos mattresses required.

Sheet steel dome casing and brass moulding on firebox back corners.

TOOLS.

A lock up toolbox built into the cab to be fitted with the

Continuation Sheet No.5. Order No. 44330

Item No. Particulars of Order and Drawing Nos.

following Tools:-

1 complete set of spanners and keys.

1 Copper hammer

1 Hand hammer

1 Coal Hammer.

2 Chisels

1 Large Oil Feeder

1 Small oil feeder

1 Hand Brush

1 set Firing Tools

1 Shovel

1 Crowbar

Toolbox to be provided with a padlock

1. 1-gall. Oil can

1 Traversing Screw Jack.

LAMPS.

1 Head and Tail Lamp 7" dia. (stock pattern)

2 Water Gauge Lamps

1 Hand Lamp

PAINTING.

Each engine to be painted Battleship Grey with grey edge, 1/4" yellow, 1.7/16" black, and 1/4" yellow, for tank and cab, but top of tank to be dark grey. Frames outside as for tank and cab, but with one yellow line only, no black. Frames inside to be red. Wheels as for tank and cab, ~~but with one yellow line only.~~ X Buffer Beams outside, vermilion,

yellow, black & claws on red.

Continuation Sheet No.6. Order No. 44330

Particulars of Order and Drawing Nos.

black edge, with white line. Inside of cab, buff with black edge and white line.

NAMEPLATES.

2 Maker's Nameplates to be fitted

Customer's Name and Number Plates as follows:-

H.H.W.D. No. 1 to 10 inclusive

PHOTOGRAPHING AND WEIGHING. added. 20.1.30.

SHIPPING MARK for Packages.

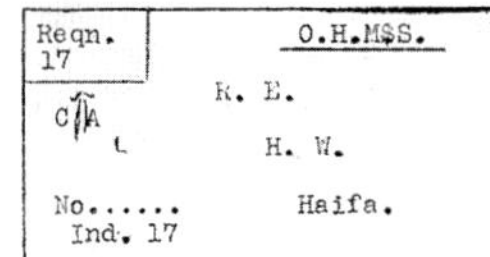

Contents of Packages.

A detailed list of contents must be enclosed when possible in each package.

MARKS.

The marks must be stencilled or painted on all articles or packages, the port of destination being in letters at least two inches high. Articles which are loose or bundled and

Continuation Sheet No.7. Order No. 44330

Particulars of Order and Drawing Nos.

are too small to bear the address are to be marked with metal labels securely fastened with wire.

In additon to the marks given in the order, the Crown Agents will furnish numbers for the packages when shipping ins tructions are applied for.

Attention is drawn to any special marking or labelling instructions that may be on the form of tender and order.

Plate 117: Showing identification as Shipton's No. 4, this locomotive was Hunslet Engine No. 1649/1930. Again, fitted with the weather curtains to keep the crew warm from the high cold winds that whistled across the quarry floor. The lettering No. 4 is a smaller version than most of the other views of this locomotive. Possible date 04/11/1961.

Courtesy Industrial Railway Society, photographer D. Clayton.

Plate 118: In steam, locomotive HE 1649/1930 stands in the sunshine. Note the thick new wooden buffer blocks obviously had only just been fitted.

Courtesy Industrial Locomotive Society, photographer J.A. Peden 23/05/1972.

Plate 119: No. 4, HE 1649/1930 static against the backdrop of the quarry face but with the Ruston excavator just seen on the right hand side of the photograph.

Courtesy Industrial Locomotive Society collection; Photographer G.Alliez 17/05/1958.

Plate 120: An earlier, larger lettering size No. 4 to that seen in Plate 117 and the locomotive looking a bit grubby.
Courtesy Industrial Locomotive Society collection; Photographer G.Alliez 11/09/1954.

Plate 121: No. 4 HE 1649/1930 in a very shabby state on the 7th of June 1959 showing the curtain pulled to protect the crew in this case from the sun! *Courtesy John Scholes collection.*

LOCOMOTIVE No. 5 – 0-6-0ST

PECKETT WORKS NUMBER 1378 NAMED WESTMINSTER BUILT 1914

Peckett & Sons locomotive works number 1378, of 1914, was constructed to class B2 under a special War Office order. The driving wheels were 3ft 7ins, with outside cylinders – 14ins x 20ins and boiler pressure 160 psi.

This locomotive, when new went to Sir John Jackson Ltd., Codford, nr Salisbury, Wiltshire and was named WESTMINSTER.

Then George Cohen, Sons & Co. Ltd. "borrowed" it and used it on the demolition contract of Larkhill Military Railway, which ran from Amesbury to the Bulford line and one other use was to propel the troop trains into Larkhill and then further on to the Stonehenge airfield.

By 1925, it was being used on the Fovant Military Railway (FMR), where a spur had been built from Dinton Station on the LSWR to the military camp at Fovant. This line made life easier for the camp supplies to be delivered and there is evidence that it was also used earlier to repatriate the wounded from the WW1 front, back for retraining.

The locomotive was acquired by the Associated Portland Cement Manufacturers Co. Ltd. (APCM) and sent to their cement factory at Houghton Regis, Dunstable in 1929.

It arrived at the Shipton-on-Cherwell site around May 1952 still with the name WESTMINSTER and became Shipton works locomotive No. 5. When redundant in 1970, it was sold, in September 1970, into preservation at the Kent and East Sussex Railway, who moved it on in July 1988 to P. Davies, Tistead Station, Hampshire, before it moved again on the 27th of October 1998 to the Northampton & Lamport Railway. Its future looks uncertain, as it is reported at the time of writing that it is up for sale yet again!

Plate 122: A fine sunshine view in June 1967, of locomotive Peckett 1378/1914 with its crew on board, again showing signs of hard wear. The wooden buffer blocks shows two very distinctive positions of the internal stone wagons (inner) and also the standard gauge buffers of the BR wagons coming in with coal and other materials needed to make the cement. *Courtesy John Scholes collection.*

Plate 123: This view of WESTMINSTER was probably taken early in its life at Shipton, as the wooden buffer plates have not been added yet and also the paintwork seems to be in quite good condition, compared with the next photograph! The brass dome has the appearance of a rifle shooting gallery back-board with numerous holes in it.

Courtesy Industrial Railway Society.

Plate 124: P 1378/1914 WESTMINSTER is seen on the 8th of June 1967 arriving from the quarry face with a full load of stone for the crusher. The lettering No. 5 can clearly be seen painted on the cab side in this image.
Courtesy Geoff Broadhurst.

Plate 125: With the high quarry wall just peeping in the background, locomotive Peckett 1378/1914 WESTMINSTER (No. 5) is standing with No. 4 (note the larger lettering No. 4) but does not look to be in steam! Note that this locomotive had the Blue Circle roundel on the top of the cab side and as far as known was the only one in the Shipton Blue Circle steam fleet to do so? Courtesy Industrial Locomotive Society, photographer J. Faithfull 24/08/1956.

Plate 126: Peckett No. 1378/1914, 0-6-0ST WESTMINSTER was designated by the Alpha Cement Co. as No. 5 in their fleet, standing here with RSH 7742/1950, No. 6 C.F.S. Just behind the locomotives, on the left hand side of the photograph can be seen the quarry face and the company houses at Bunkers Hill.

Courtesy Joe Turner.

Plate 127: A very interesting view with (in the background, right) a row of BR 16-ton open wagons standing on a quarry line behind WESTMINSTER. These could have been used for several things. Firstly, Shipton imported clinker from other cement works and there is no reason why this could not have been carried in open wagons – as also gypsum. Secondly, they also exported clean crushed limestone to other cement works, which would also have been in open wagons. Note also on the left, in the background, appears the large Rapier & Ransomes 5160 excavator.

Courtesy George Reeve of the Irwell Press, photograph by John Bonsor.

Plate 128: P 1378/1914 WESTMINSTER – works No. 5 – I expect seen here when new to Shipton works, as no internal works wooden buffer blocks have been added and it does look in pristine condition compared to later views of this locomotive. Also, no number applied or Blue Circle roundel added.

Courtesy John Scholes collection, photographer A.N. Glover.

Plate 129: This interesting front image of Shipton's No. 5 locomotive WESTMINSTER, shows how near the Bunkers Hill houses (in the distance) were not only to the works but the quarry face, seen running across the middle of the photograph.

Courtesy Industrial Locomotive Society collection; Photographer D. Stoyel.

Plate 130: A sad and forlorn locomotive WESTMINSTER, with No. 5 painted on its bunker, dumped in a siding at the Kent and East Sussex Railway's Rolvenden shed, Kent awaiting restoration.

Courtesy Industrial Locomotive Society collection; Photographer G.E. Hughes 19/7/1974.

LOCOMOTIVE No. 6 – 0-4-0ST

RSH WORKS NUMBER 7742 NAMED C.F.S. BUILT 1952

Shipton Cement Works locomotive No. 6 was built at the Newcastle works in 1952 and was delivered NEW to the APCM Ltd., Crown & Quarry works, Frindsbury, Kent and named C.F.S.

It was loaned to the Holborough cement works, Snodland, Kent in 1952 before being moved again to the Shipton-on-Cherwell cement works in January 1964 as C.F.S No. 6.

It only lasted six years and was scrapped at Shipton in January 1970.

Plate 131: A good rear view of C.F.S. No. 6, RSH 7742/1952 at rest, standing out of use on 8th of June 1967. Note the very large buffers on the rear but the front of the locomotive still has the wooden buffer blocks.
Courtesy Geoff Broadhurst.

Plate 132: An example of the livery of these works' steam locomotives. With overall apple green and the chassis, buffer beam in red and on the motion gear. No evidence of lining livery at all! The small coal bunker on the right-hand side of the cab seems to have been supplemented by storing coal on top of the firebox!
Courtesy Industrial Railway Society, photographer Gordon Green.

Courtesy Allan Baker collection.

Plate 133: A great photograph of locomotive C.F.S. No. 6 at work with its fully loaded stone train within the confines of the Shipton works. The Mitchell wagon lift dominating the photograph, plus a large pile of locomotive coal just dumped where there was a gap it seems?

Courtesy Industrial Railway Society, photographer R.K. Hateley.

Plate 134: Taken outside the small single road locomotive shed, C.F.S. No. 6 is covered in white cement. Note that Bagnall No. 2 is inside the shed over the maintenance pit that is just in view.
Courtesy Industrial Railway Society, photographer J.A. Peden 19/3/1961.

Plate 135: This photograph shows the large size of the Mitchell wagon lift at the Shipton plant, as locomotive No. 6 stands in front of it, but on the right, you can see in the distance the other end of the works giving the reader some idea of the complete area the works covered. *Courtesy the Industrial Railway Society.*

Plate 136: The grubby duo! Locomotive No. 6 C.F.S. and WESTMINSTER both taking on water in the quarry with a Bunkers Hill house peeping in on the left from the top of the quarry wall.

Courtesy John Hutchings collection.

Plate 137: RSH 7742/1952 named C.F.S., displaying its huge buffers and in steam standing outside the APCM Crown & Quarry Works, Frindsbury, Kent locomotive shed.
Courtesy Industrial Locomotive Society collection; Photographer D. Stoyel 13/04/1960.

Plate 138: Having moved on to the APCM cement works at Holborough, Snodland, Kent, locomotive RSH 7742/1952 – C.F.S., appears at rest with its chimney covered, a ladder up to the smokebox and standing on a spur near the locomotive shed. It later moved to shunt within Holborough works.
Courtesy Industrial Locomotive Society collection; Photographer G. Alliez 11/08/1954.

SHIPTON-ON-CHERWELL'S LATER REPLACEMENT DIESEL FLEET

LOCOMOTIVE – Rolls-Royce Ltd., Shrewsbury works
Sentinel number 10266 built 1967

This locomotive was a 34 ton four-wheel chain drive diesel hydraulic locomotive fitted with a Rolls-Royce C6SFL supercharged diesel engine of 255hp. The purchase price was £12,200.00 on APCM order number NWA 343/35 dated 5th of February 1968.

It was ex-works for demonstration on 4th of August 1967 at the Derbyshire Coalite & Chemical Co. Ltd., Bolsover, Derbyshire, arriving under its own power. By the 8th of November 1967 it was on demonstration at Courtaulds Ltd., Red Scar Works, Ribbleton, Preston; moving on, by the 22nd of November 1967 to Port of Preston Authority, Preston Docks.

Then travelling south, it arrived at Shipton-on-Cherwell cement works on the 12th of December 1967, again for demonstration, which was evidently successful, as it was eventually purchased by APCM hence the post-dated official order.

It remained at Shipton-on-Cherwell until May 1975 when moved to the APCM Dunbar Works, East Lothian. Note; it was subsequently recorded as scrapped or sold in 2010.

Plate 139: A fine broadside photograph of RR 10266/1967 in its works green livery whilst on trial at Shipton cement works in 1967/8. *Courtesy John Hutchings collection.*

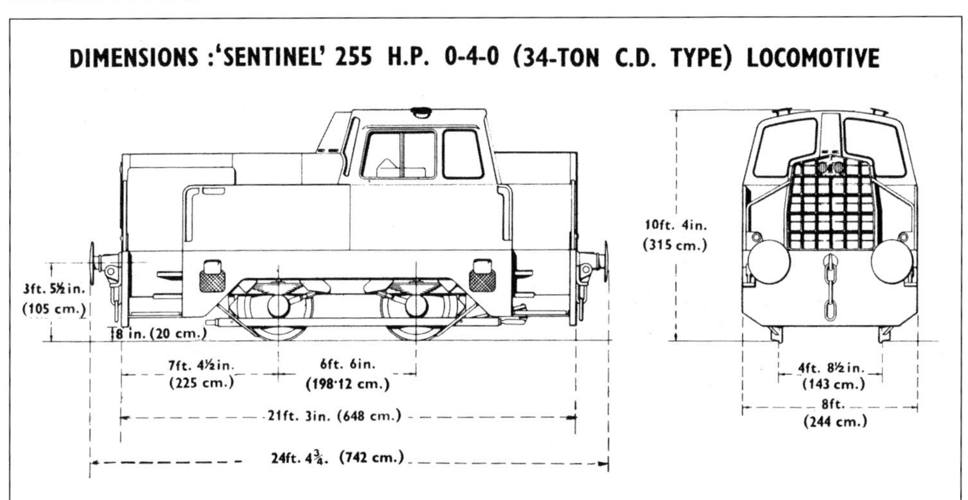

Figure 48: Dimensional drawing of the Sentinel 255hp diesel. *Authors' collection.*

Plate 140: With white splattered wagon No. 5 in tow, RR 10266/1967 in its original works green livery stands in the Shipton quarry in 1970, waiting for its crew to climb aboard to start work. *Courtesy Gordon Edgar.*

Plate 141: A visit to the APCM cement works at Dunbar, East Lothian on a sunny Saturday in July 1980 saw RR 10266/1967 parked on the end of a rake of PCA tank wagons. It was still wearing the same livery when it left Shipton-on-Cherwell works but had lost the lining detail and red buffer beam that it had new.

Courtesy Murray Liston.

LOCOMOTIVE – Thomas Hill (Rotherham) Ltd., Kilnhurst Works, Yorkshire. Locomotive TH 213V/1969. named VANGUARD No. 2

The view in *Plate 142* was taken shortly after delivery on 22nd of May 1969 to Shipton-on-Cherwell cement works. The locomotive weighed 34/35 tons, had four-wheel chain drive and was equipped with a Rolls-Royce C6 SFL supercharged diesel engine of 256hp. This particular locomotive is a later Vanguard model with roller shutter doors on the engine compartment, plus other small modifications. The cost of the locomotive was £14,047.00 and the Blue Circle Industries order number was NWA 470/43 dated 20th of March 1969.

Plate 142: The TH 213V locomotive seen here during acceptance trials working with coal trucks that arrived at the Shipton works several times a day. *Courtesy Robin Waywell collection.*

Plate 143: Probably again taken during acceptance trials, VANGUARD No. 2 locomotive is seen manoeuvring coal trucks within the works internal railway system. *Courtesy John Hutchings collection.*

Plate 144: A very fine colourful view of TH 213V/1969, shortly after delivery to the Shipton works in 1969 but still in Thomas Hill works livery.
Courtesy Gordon Edgar.

Plate 145: TH 213V/1969 stabled, operable on the 28th of May 1977 at Shipton cement works amongst the overtaking natural weed vegetation. It has been repainted in the Shipton works colour of yellow with the hand rails painted in red (although in some places well-worn!).

Courtesy Industrial Railway Society.

Plate 146: Tucked away under cover on the 9th of August 1980, TH 213V/1969 stands amongst the ever-increasing weed coverage that the works was experiencing, showing that the end at Shipton was near.
Courtesy John Scholes collection.

Plate 147: A good view of TH 213V/1969 renumbered 125 working at the Aberthaw works in Glamorgan in the late 1980s needing some loving care and a repaint.
Author's collection from international auction site and reproduced under the Creative Common licence.

Plate 148: Another look at TH 213V/1969 at Aberthaw in its last days of operation. Courtesy Adrian Nicholls.

Plate 149: The final view of the locomotive TH 213V/1969 is at the Aberthaw Works in Glamorgan (later Lafarge cement works, now Tarmac PLC), having left Shipton around March 1986.

Courtesy John Hutchings collection.

LOCOMOTIVE – John Fowler – works number 4220037 built 1966

This locomotive was an 0-4-0DH fitted with a 203hp Leyland diesel engine, new in June 1966 to APCM Sewell chalk quarry in Bedfordshire. Moved to APCM Dunstable a few years later, in February 1971, when the Sewell works closed. It was then not required at Dunstable and was sent to Shipton-on-Cherwell cement works in May 1971. See *Plate 151* on a visit to Shipton on 28th of May 1977 parked at the coal tippler. It was seen workable again in 1982 but it had left Shipton by about 1986.

Plate 150: An unusual back view of JF 4220037/1966 standing within the works railway system on the 9th of August 1980.
Courtesy John Scholes collection.

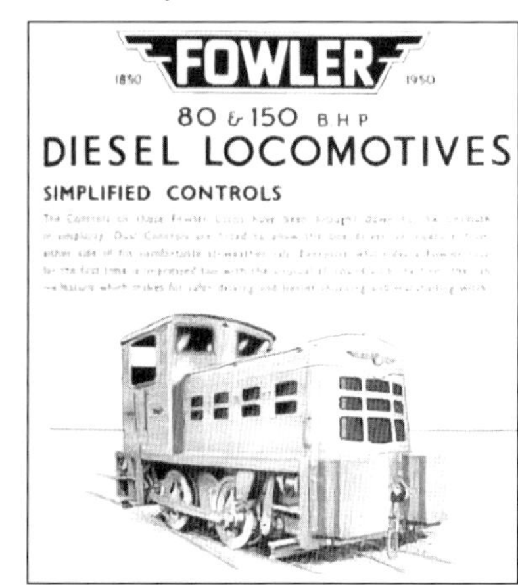

Plate 151: This view is JF 4220037/1966 when the locomotive was at Shipton cement works in Blue Circle days. Note the two extra air filters fitted on the footplate to reduce dust intake to the engine.
Courtesy Industrial Railway Society.

Plate 152: A final look at diesel JF 4220037/1966 on 28th of May 1977 standing on one of the disused internal lines in the plant. Courtesy John Scholes collection.

Rail operation in the Shipton quarry is thought to have ceased in 1972-73, when the dump trucks took over moving all the stone from the quarry face to the crusher. The quarry rail track was lifted in 1974.

An IRS members' report dated 30/1/1972 reported JF 4220037/1966 seen working in the quarry, TH 213V/1969 stabled in the middle of the factory and RR 10266/1967 inside the locomotive shed.

An appropriate time to mention the Industrial Railway Society and the Industrial Locomotive Society. Industrial Railway Society members go out to industrial sites all over the country and send in reports to the various area handbook editors, so that records can be and are kept up-to-date on locomotives and sites for the new editions of the reference books that are published every so often. The IRS was founded in 1949, so when this book is published it will be in their 75th year.

It was founded as the Birmingham Locomotive Club Industrial Locomotive Information Section. One of its aims was to cater for those interested in privately owned locomotives and railways. The membership receives a quarterly 48-page colour magazine, with over 250 issues already published over the years and in addition it is one of the leading publishers in the industrial transport field with many interesting, and superbly produced lavish books. The society has just opened a specially built archive research centre at Statfold Barn where researchers can visit by arrangement to see the many plans, photographs and comprehensive library that the society has amassed.

The following interesting text is taken from a report by an IRS member regarding a visit to the Shipton-on-Cherwell cement works on the 27th of October 1973 and included verbatim:

> **A.P.C.M. Oxford Works, SHIPTON-ON-Cherwell. 27.10.73**
>
> Since Flying Scotsman was operating between Tyseley and Didcot on this date, it seemed a good opportunity for a visit here; however, the works manager's reply to my letter for a visit was a refusal, on the grounds that "*we have now converted from rail to road dumper haulage*".
>
> However, in passing the works it was noted that there was rail action in the quarry, and a little negotiation resulted in being allowed entry to the quarry (unofficially), helped by the manager being absent that morning and the foreman saying "if anyone asks you, you are looking for fossils, along with the other two chaps in there already" (apparently genuine geological types). I therefore obtained some shots of RR 10266 at work in the quarry, working at both the 'hard' and 'soft' quarry faces (i.e. stone and clay) and then seeing in action the tippler, which bears a striking resemblance to a now obsolete BR type of tall and huge coaling plant, with wagons being lifted bodily some 40-50 feet. Spare locomotive Vanguard TH 213V in yellow, whilst the Sentinel still being green. The Fowler was at work with coal wagons from BR and was in yellow, JF 4220037. The Sentinel's livery was rather mottled, probably the original paint still, now worn down to the undercoat! However, the loco crew were of course more interested in Flying Scotsman, which passed in fine form.

The above is a very good example of the reports that society members send in to update records and has given the readers positive first-hand information of the work at the Shipton quarry, plus the state of all the diesel fleet that were still there, albeit that the works manager had implied there was no rail operation. It also confirms that the rail track was still down on the quarry floor in October 1973 and rail use much still in evidence.

A further member's report dated the 22nd of March 1975 came in from another member regarding the locomotive status at Shipton and was as follows:

> DGN1 (RR 10266) was out of use in a quarry shed and isolated from any of the remaining lines.
>
> SGN 1 (TH 213V) was still a working locomotive.
>
> SGN 2 (JF 4220037) was out of use but was the standby locomotive.
>
> No rail traffic remains in the quarry now, the only activity being in the exchange sidings and around the factory area. Only one locomotive is required for work, the JF being the standby, but even this looks as though it has not worked for many weeks (or months!). Only one quarry wagon was seen and the only traffic appears to be coal from BR for the works.

SHIPTON CEMENT WORKS INTERNAL STANDARD GAUGE WAGONS

The majority of wagons working into the Shipton-on-Cherwell cement works were standard main line types but in the quarry purpose-built wagons were used. In the late 1920s, many modernised cement works, often as part the replacement of old narrow gauge quarry railways by higher-capacity standard gauge systems, were equipped with secondhand standard main line wagons, mostly with wooden bodies; these were far from robust and in almost all cases were soon replaced with stronger steel-bodied wagons. Shipton cement works, quarrying a hard and abrasive rock and an entirely new site, was equipped with purpose designed and built standard gauge steel wagons from the start.

There is little direct contemporary evidence, but it seems fairly clear that the original design had bodies with a low-sided, wide V cross section and solid disc wheels with inside bearings and generally worked in trains of five wagons. Photographic evidence is that the first and fifth of each 'set' of wagons was braked and the intermediate three unbraked. The brakes were only 'parking' brakes, to hold the wagons still during loading or unloading. The design was matched to the original tip into the primary crusher and their capacity was apparently around 8-10 tons. Little else is known about them and deduced only from photographic evidence, in particular the method of tipping.

There are two points that seem relevant. First, on one side of the body was welded some steel eyes connected with a steel bar. A likely explanation is that, at the tip, this engaged with either an overhead chain/trolley hoist, or a floor-mounted hydraulic strut, by either of which the body could be raised to a tipping position. Second, at the four corners of the frames are eyes to which the body could be pinned. Tipping would thus consist of propelling the train into position for the first wagon to be tipped, removing two safety pins on the opposite side, pinning down one side of the chassis, engaging the hoist and raising and tipping the body. The locomotive would then move the train along one wagon and repeat. That was not an uncommon method of unloading rail wagons. Because the body could, in principle, be tipped either side, it did not matter which way round the wagons were, as they were probably the same on each side. However, perhaps to increase capacity, some or all of the wagons were modified by the addition of a short vertical steel sheet on one side only (see Plate 153), the wagons thus became asymmetrical and would have had always to face the right way. Also, if this was the mode of tipping, then it rather negates the wagon brakes, as a locomotive would have remained attached throughout tipping the train.

Plate 153: The original wagons with the low sided wide Vee-body, as later modified by the addition of a steel extension piece on one side only. This was probably added to try and gain an increase in stone capacity. The photograph also shows that the front wagon has a bed formed of RSJs, probably used to carry heavy items around the quarry, for example a navvy bucket for repair. *Courtesy Industrial Railway Society.*

Plate 154 (above) and Plate 155 (below): Two views of No. 3 (AB 2041/1937) with a loaded rake of stone in the original quarry wagons. *Courtesy Lawrence Waters, Great Western Society Trust collection, Didcot.*

The drawing in Figure 49 shows the later design of quarry wagon, with a larger, more square form of body having roughly 50% more capacity. These wagons were the predominant design in the later days of the works operations, although the older original wide V-section wagons were never replaced and still used. This was because the new Mitchell hoist tippler was an addition to, not a replacement for the original tip. The new hoist could take wagons of larger capacity, but it could not handle the width of the original design, and therefore both continued in use. These larger wagons were also operated in rakes of four or five as they were all individually braked not like the original version.

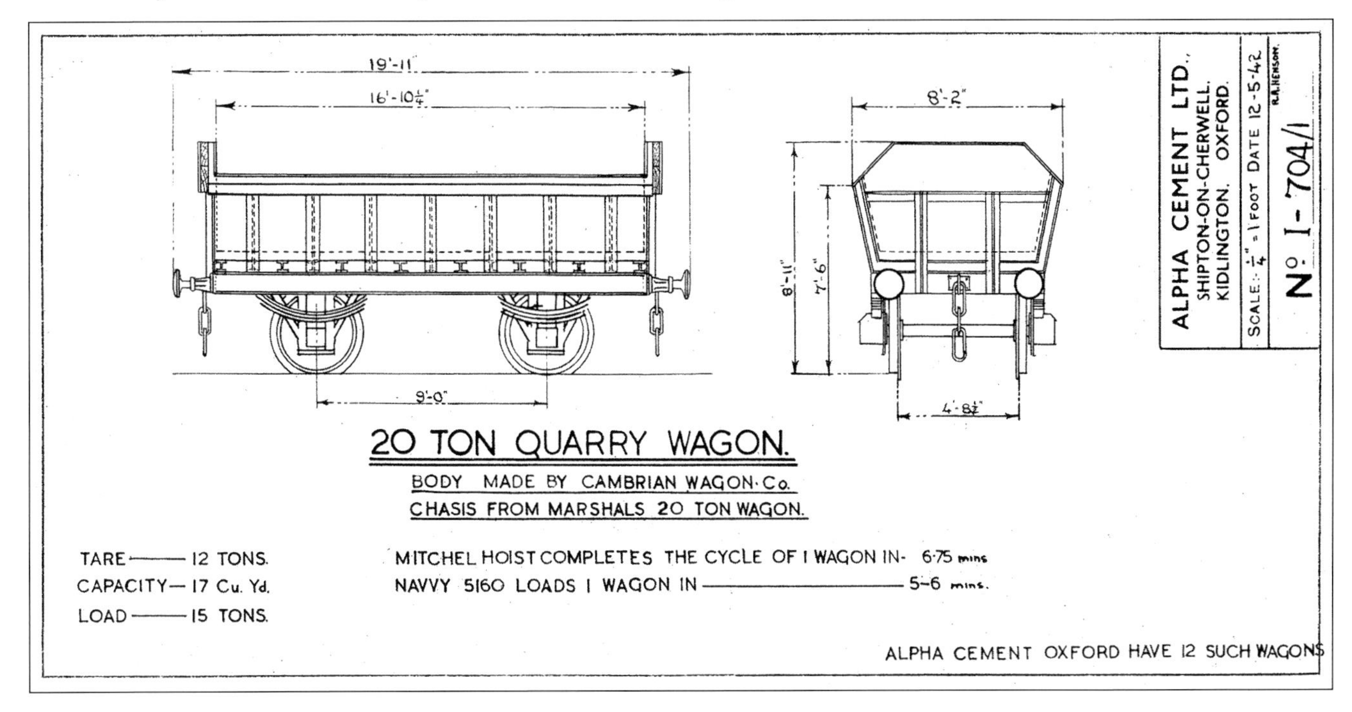

Figure 49: This wagon drawing by R.R. Henson is dated 12th of May 1942 and shows the image of the wagons on the train in Plate 156 with the axle boxes on the outside of the wheels. It states that Alpha Cement had 12 of these wagons and notes the interesting facts on the drawing that the quarry navvy excavator is listed as 5160 type made by Ransomes & Rapier and the cycle times it took to fill one wagon load was 5-6 minutes. Then, when the wagon load was transported along to the wagon Mitchell hoist, the time to hoist up, tip and then drop down was 6¾ minutes. One feature not shown is that every one of these wagons was braked. Courtesy Chris Down collection.

Plate 156: This view illustrates both styles of the internal wagons being used at the same time in the Shipton quarry. AB2041/1937 is coupled to the loaded, fully-braked outside axle box more modern wagons, whilst the four outside wheeled version sits alongside, also full with quarry stone. The different 'vee' sided shape comparison of the two wagon types is portrayed extremely well in this photograph. Courtesy Joe Turner.

Plate 157: The final form of the internal stone wagons that were bigger and had the axle boxes on the outside of the frames, as per the drawing in Figure 49. Courtesy Laurence Waters, Great Western Trust collection, Didcot.

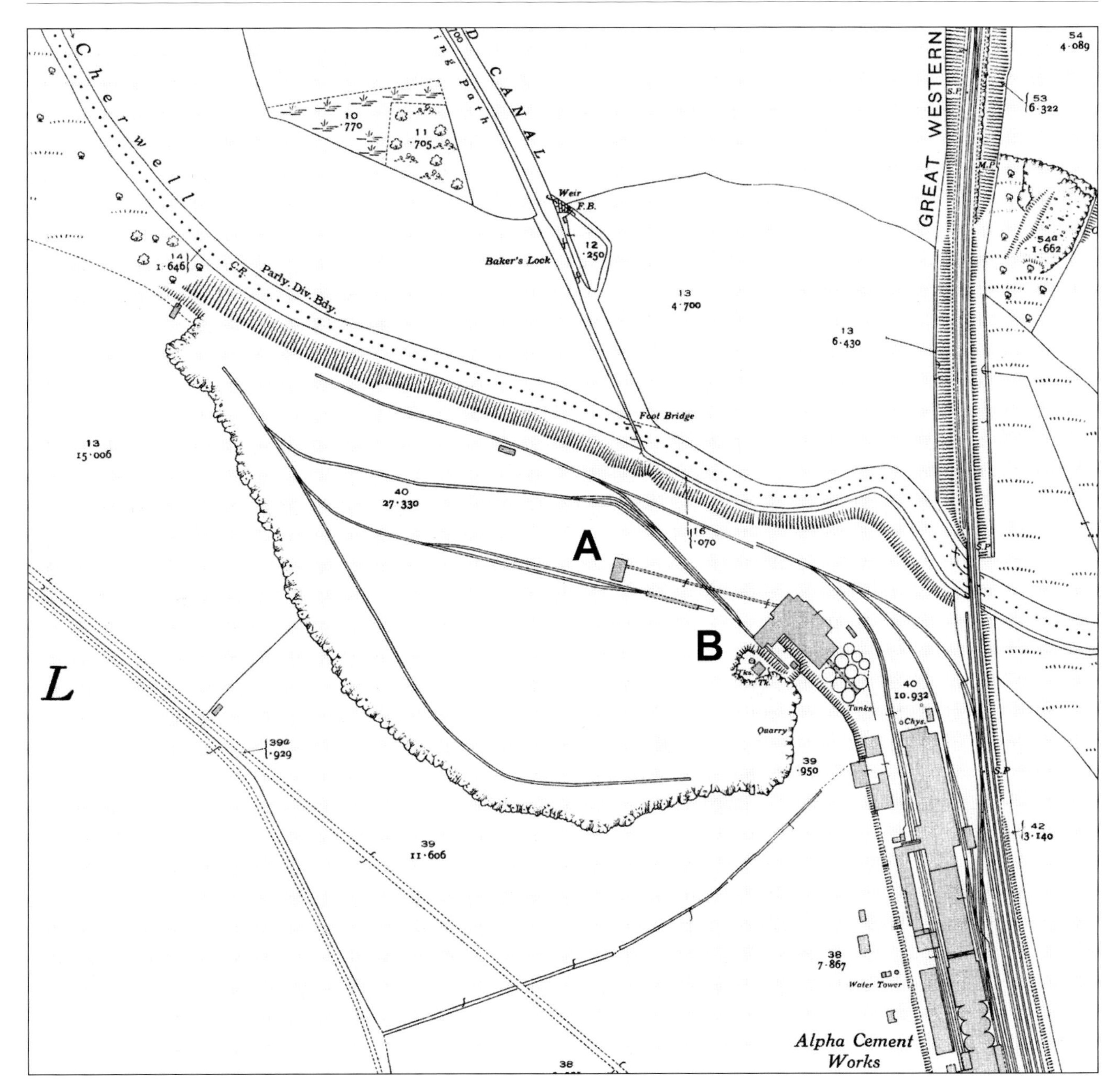

Figure 50: Portions of two OS sheets, revised in 1937 and published in 1938-39, joined together to show the two raw material tipplers. 'A' is the original installation, and 'B' position the additional Mitchell wagon hoist added about 1936. As can be seen, the rail tracks were so arranged that trains of rock could serve either tippler, according to which rake of wagons they had and what were the requirements needed.

Reproduced courtesy National Library of Scotland under the Creative Commons Attribution Licence.

When, from about 1970-71, rail haulage in the quarry began to be replaced by dump trucks, the original tip ('A') was replaced by a new tipping point located further into the quarry, to receive stone unloaded end-on by dumpers. From there a new conveyor fed the original tip and then, via the original conveyor, up to the top of the now redundant Mitchell tippler.

STANDARD GAUGE MAIN LINE WAGONS USED FOR CEMENT TRAFFIC

This section touches on just one type of cement wagons, used both by private companies and by British Railways – the PRESFLO.

At the time of writing no evidence has come forward regarding the use of any other type of specialist cement wagon by the Blue Circle at Shipton-on-Cherwell works.

Plate 158: The above photograph illustrates the Blue Circle Presflo wagons, showing the two alternative styles of Blue Circle Company branding advertising on the wagons. The photograph is dated 11th August 1966 and was taken at the Blue Circle Westbury Works, Wiltshire. These wagons are identical to those used across the Blue Circle works and possibly worked from the Shipton cement plant. *Courtesy Chris Down collection.*

Not until the 1950s did a market in bulk cement emerge that required specialist road and rail vehicles. Prior to then, cement was typically despatched in bags loaded into railway company vans. The 35 ton capacity Presflo wagon was developed by British Railways to cater for this traffic. The name was a shortened version of 'Pressure Flow;' itself an abbreviation of the rather lengthy 'Pressure Discharge Bulk Powder Wagon'. These wagons were probably used for bulk cement despatches from Shipton-on-Cherwell works, certainly APCM/Blue Circle owned a fleet of these wagons, painted grey to differentiate them from the brown or bauxite colour used by British Railways. The wagons were top loaded under gravity from a cement silo, but emptied by air pressure through a flexible pipe operating valves on one side of the wagon either into a storage silo or a road vehicle.

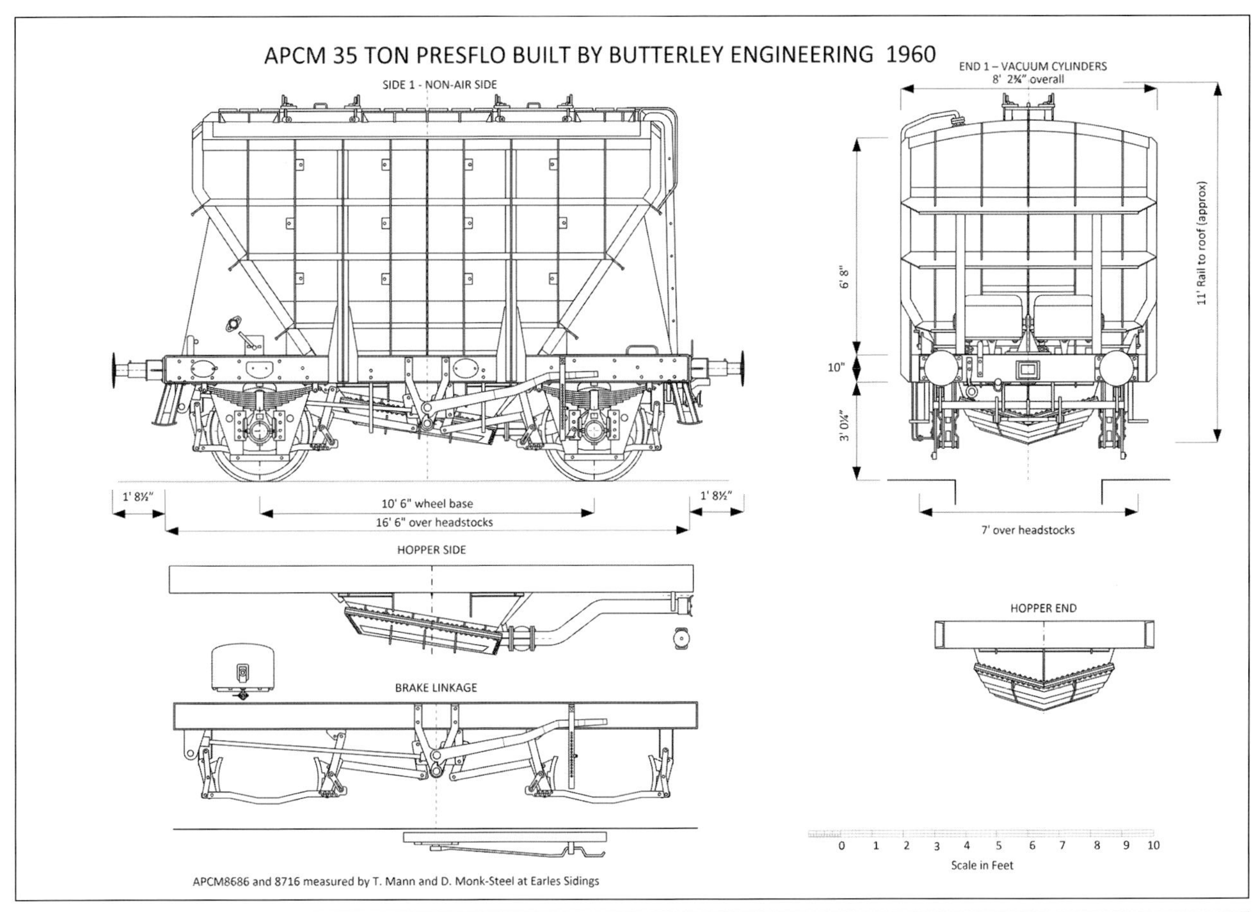

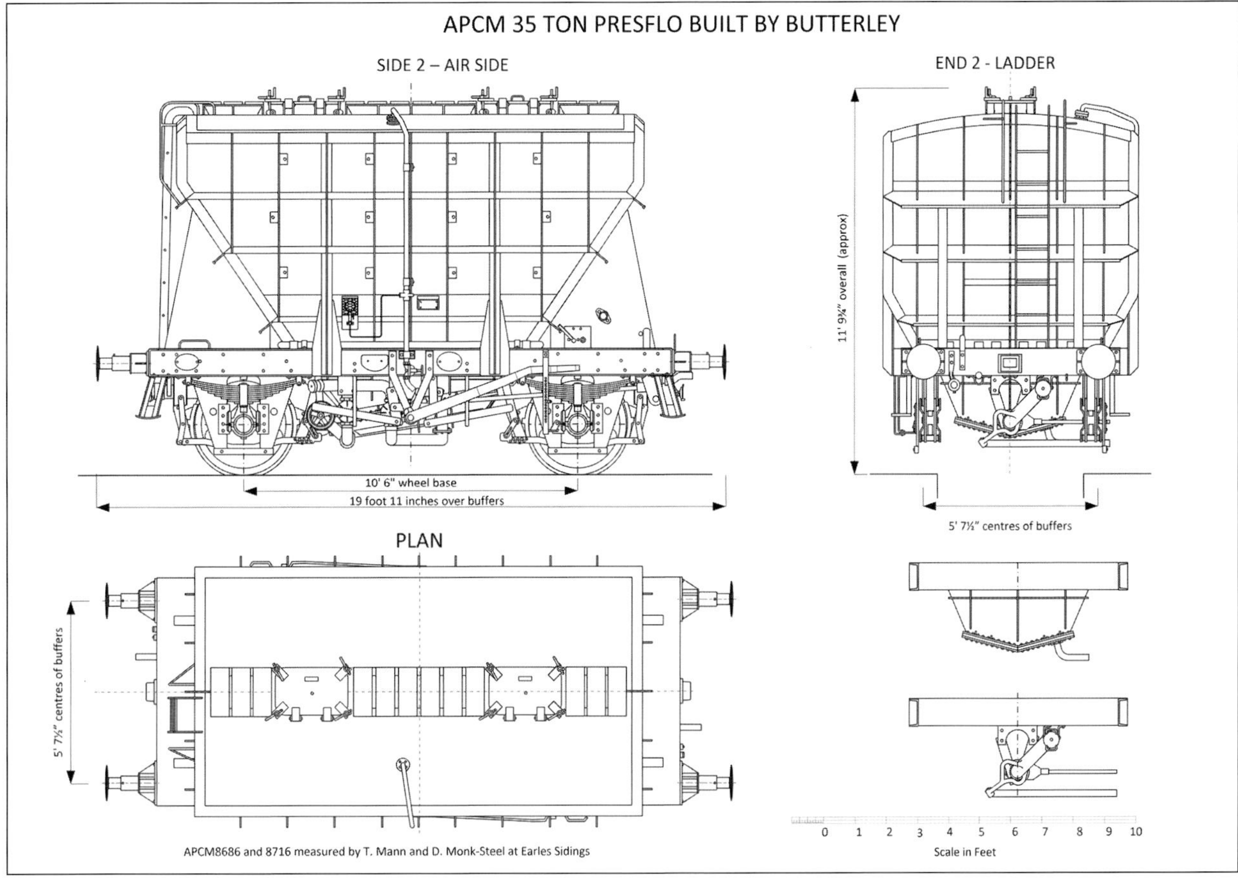

Figure 51: These two superb drawings by T Mann and D. Monk-Steel show the APCM 35-ton Presflo wagons that might have been seen and loaded with cement at the Shipton works. Courtesy of the Artists involved.

By kind permission of the *Great Western Railway Journal* of Winter 2012, an extract is included from the GWR Reading goods sidings article that proves that Presflo wagons were used by Shipton Cement works and reads:

> Local directories for the GWR Reading goods yard show that the Oxford Shipton Cement Company was there in 1931, and in 1949 there is mention of the Alpha Cement Company. Certainly, by the 1950s the Cement Marketing Company were receiving supplies in bulk pumped from Presflo wagons into lorries and this took place in the next siding over from the Co-op siding. Peter Hutt recalls *"It was surprising how hot the cement was when it came out of the wagon. Sometimes some spilt onto the ground and I remember scooping it up and it was quite hot I can tell you."*

The first batch of 200 BR Presflo wagons were constructed during 1954 and had a single vacuum brake cylinder and longer hand brake lever. The later models and more common type had twin vacuum brake cylinders and a shorter brake lever.

APCM eventually bought 228 of their own PRESFLO wagons numbered PF01 to PF102 (TOPS numbers APCM 8601 – 8702) from Butterley Engineering in 1960, and PF103 to PF162 (TOPS numbers APCM 8703 – 8762) also from Butterley in 1963, and PF163 – PF228 (TOPS numbers APCM8763 – 8828) from Metro Cammell Co.in 1963.

PF01 to PF102 were at first painted yellow, nearer to a bright primary yellow; unfortunately, they soon became contaminated with cement dust and, as anyone who has lived close to a cement works in the 1950s will confirm, the dust clings to any surface and stubbornly resists cleaning. The construction of the Presflo wagons did not help either with outside framing giving a lot of corners to conceal dirt. APCM very quickly realised that they were wasting their time and changed the livery specification to grey all over. The bottom section of the two side stanchions were often picked out in yellow and was presumed to highlight a potential hazard, being a projection at or near head height for anyone working alongside a wagon.

All the APCM owned wagons originally had name boards attached. These were large yellow rectangular boards, much as in the Airfix kit, but later these either fell off or were removed. The welded brackets however remained visible throughout the life of the fleet.

British Railways Presflos hired to APCM had a 'target' logo of a blue circle with a horizontal bar having the firm's name applied on a yellow rectangle. A few APCM wagons also got this embellishment. These too dropped off over time.

The body configuration of the APCM wagons changed between batches depending upon which firm built them. The angle between the vertical surfaces and hopper slope of Butterley built wagons PF01 to PF162 was a hard fold line. But the Metro Cammell version PF163 to PF228 this join was a smooth curve. There are other small differences between the BR and APCM vehicle, including the pipework, and arrangement of number plates and instruction inscription.

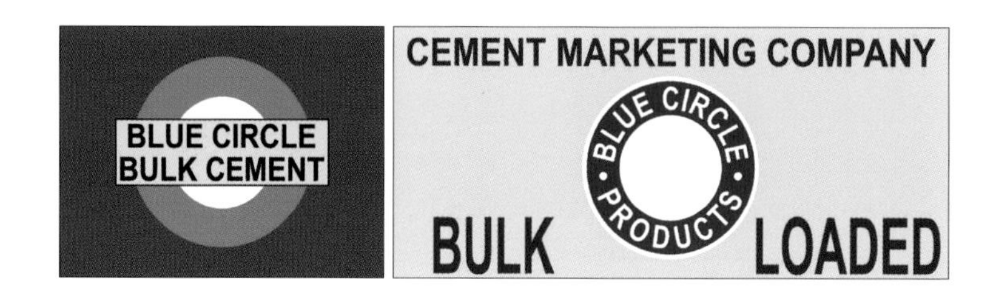

Chapter Nine

The demise of the SHIPTON CEMENT WORKS

As part of the preparation for the projected planning application for the new works, due to be submitted in January 1983, a feasibility study was carried out on the future management of the huge 'Northern Lake' that had formed in the worked-out portion of the quarry, as a bird and plant refuge, until the smaller lake at the southern end of the quarry was available.

However, there was a significant problem in that the new works proposal included a raw materials conveyor which would pass right through the northern lake, not to the north of it as originally thought. This meant that the lake could not become the proposed wildlife refuge.

One solution put forward was to construct two bunds across the lake, the northern one becoming the foundations for the raw materials conveyor and the southern one forming the new limit of the lake. Plants could then be relocated into the somewhat smaller wildlife habitat and remain undisturbed well into the projected 32-year life of the new works. If planning permission had been granted, this would have happened over the winter of 1983-84.

Of course, the new works never happened and the actual date of closure was only a few years away. Nevertheless, this does demonstrate the thought and planning that Blue Circle were prepared to put into their environmental responsibilities.

The actual closure of the Shipton-on-Cherwell cement works took part in two separate stages. The production of cement from clinker made at Shipton seems to have ceased in October 1986 when the local newspaper stated that 120 workers were made redundant out of a works force estimated at 150. The remaining 30 or so workers were kept on by Blue Circle Cement to keep parts of the plant in use, particularly for grinding clinker imported from other cement works, until 1993, after which the gates were shut for good. Not until February 1996, did a consortium (Drexfine Holdings), who had previously purchased the derelict works, apply for planning to turn the site into a waste disposal plant – without success!

After the final closure in 1993, environmental specialists descended on the site before it had even been sold. They studied the returning plant and bird life, with over 52 species of birds recorded. The site was on track to becoming a Site of Special Scientific Interest (SSSI).

Details of the legal owners/or lessees of this huge site, from its closure are a bit vague. Drexfine Holdings Ltd. became involved in 1994, with a subsidiary, Woodstock Leisure Ltd. Drexfine put in several planning applications; one that was passed being entitled: *Phase One;* to carry out perimeter stabilisation work. Then construction of earth berms to include the deposit of imported inert waste to prevent collapse of the roadway, which was permitted in October 1995. A further application for permission to stabilise more areas of the perimeter was requested and permitted.

In 1997, Bride Parks Ltd. (later the Kilmartin Group) purchased the quarry site and the full story of this company's efforts to get various planning permissions has been kindly written by the company's directors for this book and is included in full in *Appendix Three*. Eventually they sold the site to Earthline Ltd. in 2010. However, a long list of planning applications since then exists, with to date, no positive solution to the future of this huge quarry.

It is worth mentioning one such proposal submitted to the Planning and Regulation Committee on the 29th of October 2018, by a company called Shipton Ltd. of Chester, Cheshire, proposing extraction of mineral and restoration by infilling with imported inert materials to agriculture on land to the south east of Shipton-on-Cherwell. It stated that the quarry was situated 6.2 miles north-west of Oxford, near Bunkers

Hill, was in the Green Belt and was designated an SSSI of geological importance. Also as a Conservation Target Area (CTA) and an important designated County Wildlife Site – still mentioning the vague Long Barrow that seems not to exist!

The application was never passed. Earthline Ltd. was contacted but never answered the request for information on the future of the Shipton quarry.

Plate 159: An impressive view photographed in the late 1980s of the Shipton-on-Cherwell cement works situated in Oxfordshire. It was photographed from alongside the ex GWR Oxford to Birmingham main line and shows that all the exchange sidings had been removed and the vegetation had already taken over the track beds. This view was certainly captured just before final closure of the plant.

Authors collection acquired on an international auction site.

Plate 160: This view of the quarry floor, looking northeast from the works access road of the plant, as it appeared towards the end of its working life (the date is after the lifting of the quarry railway tracks). The process to cleaning up the whole area had started to take shape after the works had reduced cement production.
Courtesy Oxfordshire County Council, Oxford History Centre; ref POX0158148.

Plate 161: Shipton-on-Cherwell quarry viewed looking northeast on the 7th of June 2015, photographed by Mat Fascione and is reproduced under the Creative Commons License. Only the tall white chimney remains (just a couple of months before that was demolished, see Plate 164) of the Shipton cement works with the vegetation and flora starting to take hold of the quarry once more.

Plate 162: A superb aerial view taken in the 1990s for the Kilmartin Group, covering the whole Shipton-on-Cherwell quarry and works area, which shows the quarry encroaching very near to the Bunkers Hill housing site (top right) – the old GWR Woodstock branch line track bed is very evident, snaking from the bottom left up through to the top left. The main railway lines from Oxford to Banbury running from bottom left to middle right and lastly the edge of Shipton-on-Cherwell village on the left.

Courtesy Kilmartin Group Company.

Plate 163: A superb photograph of Class 58, No. 58002 DAW MILL COLLIERY passing the derelict remains of the Shipton-on-Cherwell cement works with the 7M28, 19.00, Oxford Hinksey Yard to Bescot Departmental Yard on the 30th of July 1999. Note the locomotive is still carrying the name of the 1Z58 THE WORKSOP ABERDONIAN on the front from a rail tour that took place some years before! The middle ground, with colourful weeds and bushes, was the area of the track beds of the old three exchange sidings.
Courtesy Martin Loader collection.

Plate 164: Earthline Limited were the demolition contractors assigned to demolishing the old Shipton-on-Cherwell cement works and one of the last structures to be demolished was the 250-foot white chimney, seen here on its way down, in front of a large crowd of local photographers on the 30th of August 2015.
Photograph courtesy the late Ted Thornton.

Appendix One

KIRTLINGTON CEMENT WORKS – Annual OUTPUT production, details of WAGES and EMPLOYEES TIME LOST.

These figures are reproduced from the Journal of History and Technology, Volume 9, Number 3, 1972

ANNUAL PRODUCTION OF CEMENT AT KIRTLINGTON, 1907-1922

Year	*Cement made in tons*	*Cement sent out in tons*
1907	481	481
1908	3802	3680
1909	6293	3936
1910	3997	4181
1911	8464	7984
1912	8468	8383
1913	11,568	11,299
1914	13,910	14,072
1915	13,335	13,067
1916	11,803	11,375
1917	12,204	13,577
1918	11,764	11,421
1919	11,996	12,279
1920	12,978	12,835
1921	12,810	11,724
1922	16,673	17,287

DETAILS OF THE WAGES PAID AT KIRTLINGTON CEMENT WORKS

	1916	
Under Manager	£3	per week
Foreman	£2	" "
Quarry Foreman	28s	" "
Mechanics	7d	per hour
Mechanic's Mate	4½d	" "
Drivers Power plan	7½d	" "
Stokers	6½d	" "
Head Millers	7½d	" "
Millers	6½d	" "
Quarrymen	6d & 5½d	" "
Putters in Foremen	6½d	" "
Putters in	6d	" "
Packing Foremen	6½d	" "
Station Foremen	6d *(plus 2s bonus)*	
Works Foremen	11d*	" "

	1925	
Foreman Mechanic	£4 7s*	per week
Mechanic	1s 1d	per hour
Carpenter	1s 1d	" "
Foden Drivers	1s 2¾d	" "
Foden Mates	11d	" "
Drivers	11d*	" "
Stokers	10d*	" "
Burners	1s 1d*	" "
Millers Chief	11d*	" "
Millers Ordinary	9½d*	" "
Cement packers ?		
Cement Foremen	11d	" "
Sack repairer	5s 4½d per 100 sacks	

Packers and Station				*Station Un-loaders*			
Men	5½d	"	"	Men	10d*	"	"
Foremen	11d*	"	"				
Shift men	6d	"	"	Crane Driver	10½d*	"	"
Quarry Foremen	1s 1½d	"	"				
Quarrymen	10d	"	"				
Quarry Truckmen	10d*	"	"				
Crushers Chief	10½d*	"	"				
Crushers Ordinary	10d*	"	"				
Chaff Cutter	30s*	per week					

*Note: * Means – plus bonuses.*

Note: From 1925 to 1929, a bonus of 1d per ton was paid on the cement made over 300 tons per week

TIME LOST ON THE KILN AT KIRTLINGTON, 1919-1927 DUE TO ALL CAUSES

Days	***Stopped***	***Reason***
1919	57	4 days due to coal strike plus 4 holidays
1920	58	
1921	107	81 days of coal strike
1922	39	
1923	23	
1924	80	31 days due to Builders strike

Appendix Two

MAINLINE trains passing SHIPTON-on-CHERWELL CEMENT WORKS.

Plate 165: A scene looking north next to the APCM Shipton-on-Cherwell cement works, on the left showing the two remaining holding sidings of the cement works next to the ex-GWR main line. The line on the far left is the entry into the cement works controlled at this time by a ground frame situated behind the locomotive on the right. This Reading shedded locomotive, No. 6924 GRANTLEY HALL drifts down the slight gradient with a class 8 express freight and the lowest amount of fitted wagons. The ex-GWR signals are already 'off' for the expected north bound train on the 29th of July 1965.

Courtesy Steve Banks collection, photographer R. Brough from the R.K. Blencowe Archives.

Plate 166: Bereft of shed and nameplates, Oxford's shedded Hall class, No. 6910 GOSSINGTON HALL is not the prettiest of sights. On a down parcels train, with a clear road ahead, it coasts past the APCM Shipton-on-Cherwell cement works (seen on the right) on the 29th of July 1965.

Courtesy Steve Banks collection, photographer R.K. Blencowe.

Plate 167: This early 1980s view looking north towards Banbury is particularly interesting as it shows the three exchange sidings, with coal wagons on one of them. The Shipton cement works position relative to the ex GWR main line shows up well with this double headed Crompton diesels hauling a freight train, passing the cement plant on its way (via Oxford) to the south. *Courtesy Ian Cuthbertson.*

Plate 168: A Merry Go Round train from Didcot Power Station taken on 22nd of May 1982. Locomotive Class 47, No. 47 360 is seen approaching the Bletchington road bridge with coal empties from the Power Station to Mantle Lane.

Courtesy Martin Loader collection.

Plate 169: After the demise of the Speedlink network in early July 1991, MoD traffic from the south of England made its way north, via the 6M79, 15.08, Eastleigh to Crewe, Basford Hall working. This train would often have multiple locomotives on the front, as it was means of getting surplus motive power north. On 29th of July 1991, this train was seen passing the old Shipton-on-Cherwell cement works powered by class 37, No. 37 235, 47 280 PEDIGREE and 47 851.

Courtesy Martin Loader collection.

Appendix Three

SHIPTON-on-CHERWELL QUARRY ownership 1997 – 2010 KILMARTIN GROUP – A Company Statement!

SHIPTON-on-CHERWELL QUARRY, OXFORD
Project Description

Introduction and Background

Shipton quarry lies 6 miles to the north of Oxford, bounded by the A4260 Banbury Road to the west, the Didcot – Banbury main line railway to the east, the River Cherwell/Oxford Canal to the north and the disused track bed of the former Blenheim and Woodstock Branch (B & WB) line to the south. The 168-acre site formerly housed a quarry for the winning of limestone and a cement works fed by a rail connection to the main rail line. An adjacent 240 acres was acquired which covers ground subject of further consented limestone quarrying. The site ceased quarrying in the 1980s and cement production stopped in 1993. Since that time, it lay dormant except for some stabilisation work to the quarry walls.

Bride Park's principals were originally involved in the site as a Regional Distribution Centre for BMW Rover finished vehicles, winning Planning Consent on appeal for a 60-acre rail connected facility in 1999. The site was subsequently purchased by Bride in 2001 following the break-up of the BMW Rover Group and their withdrawal from the project.

Following a review of the site's potential carried out by RPS Chapman Warren, a specialist planning practice, Bride pursued the allocation of the site as a new settlement through the Structure Plan process.

Planning History

There is documentary evidence that the site was used for the extraction of minerals as far back as the 1930s. However, the first planning permission granted for the quarry was an Interim Development Order (IDO) in June 1948. Interim Development Orders were effectively planning permissions granted between 1943 and 1948, which were subsequently preserved by successive Planning Acts as valid consents either where development of land had not commenced or not been completed by the 1st July 1948. Many IDOs were granted during this period with few planning conditions, as the priority was to ensure an adequate supply of materials for post war reconstruction. After this there are several separate planning applications made by the operators of the quarry, one in 1955, one in 1977 and two in 1983 before extraction operations ceased in the late 1980s/early 1990s. Except for a small area of only 1.6 hectares immediately adjacent to the village of Shipton-on-Cherwell, none of these consents appear to carry restoration conditions; indeed, no restoration work has ever been carried out within the quarry area (but see below as far as stabilisation is concerned) and it remains as something of an eyesore within the landscape.

60-acre Vehicle Storage Permission

In 1997 when the quarry changed hands, the new owners submitted a planning application to Cherwell District Council for the use of the northern part of the quarry floor and the former cement works to the east for the storage of motor vehicles for a 10-year period. This application included demolition of the existing derelict buildings together with further stabilisation of the quarry face. The District Council refused planning permission on the grounds that it represented inappropriate development within the Green Belt and that special circumstances advanced by the applicant did not outweigh the presumption against development. The application was subject of an appeal to the Secretary of State who granted permission for the car storage for a temporary ten year period from 1998 onwards, subject to several planning conditions, including major restoration works. This consent has never been implemented although a renewal was granted permission by Cherwell District Council in July 2005.

Cherwell Local Plan

The Deposit Draft and the subsequent revised Deposit Draft had a site-specific policy for the quarry which stated "replacement footprint, buildings of reduced height to the current cement works and no greater traffic levels than the previous use (cement works)" would be possible. Delays in the programme for the Local Plan caused Cherwell to abandon the plan and move forward with a new Local Development Framework.

The draft policy to the GB7 application relating to Shipton quarry stated:

Proposals for the redevelopment of the worked area of Shipton-on-Cherwell quarry and the cement works, identified in this plan as a major developed site in the green belt, will not be considered inappropriate development provided they are set within the context of a comprehensive plan for the whole site and meet the following criteria:

1. The proposals would have no greater impact than the existing development on the openness of the green belt;
2. The proposals would contribute to the achievement of the objectives for the use of land in green belts;
3. The proposals would secure acceptable environmental improvement of the major developed site;
4. The proposals would include the removal of all existing plant and buildings on the whole of the major developed site;
5. The traffic and travel implications of the proposals are acceptable, considering the previous levels of employment and traffic generated by the site;
6. The amount of new building floorspace would not occupy a larger area of the site than the existing building footprint;
7. The proposals maximise the opportunity to make use of the railhead;
8. the proposals would not result in damage to or loss of the site of special scientific interest;
9. The proposals would not injure the visual amenity of the green belt and (where new buildings are proposed) will include a significant reduction in height compared to the existing cement works; and
10. No development is constructed below the level which is likely to flood naturally;
11. The proposals would not have an adverse impact on the Oxford canal.
12. Consent granted for redevelopment will include a requirement that any existing planning permission applying to the site be revoked without compensation.

Delays in the programme for the Local Plan have caused the District Council to abandon the formal procedures before the policies were tested at Public Inquiry.

Brownfield Status

The site was acknowledged as a "brownfield site" by the Minister in the appeal decision for car storage and subsequently by Oxford County in a letter.

PPG 3 Housing

The Government required planning authorities to focus attention on brownfield sites for new housing. PPG 3 contained a well-established search sequence and the site met the criteria generally.

Structure Plan Panels Report Dec 2004 and Emerging Regional Spatial Strategy to 2016(RSS)

The panel confirmed that the site was the only major brownfield site in the green belt and it should not be ruled out as unsustainable. The opportunity to link the site to the rail line was recognised by the panel. After the Planning Act 2008 the emerging RSS had to take account of the review by the Chancellor to increase the land supply for housing.

Planning Summary

Bride Parks established a substantial project team and carried out numerous studies covering planning, transport, environmental and services reports over a three year period.

Bride also carried out a series of discussions, negotiations with County, District and Parish Councils along with relevant statutory bodies and local interest groups.

Stabilisation work

The site produced income from 1997 from the stabilisation activity involving the importing of inert material such as waste from demolition. Planning Permissions were granted in several phases by the County Council who are the Minerals and Waste Authority for the site.

A contract was entered into with a specialist earthworks contractor called Earthline Ltd. Their work was supervised by another local firm called Ablegrove Ltd. who monitored the material entering the site, organised the invoicing of the loads based on tonnage and looked after general security.

The last phase of the work was completed and further consents to continue the stabilisation work were considered.

A new joint venture company between Bride Parks and Kilmartin Group was set up to acquire the freehold site and adjacent option from Bride Parks (Oxford) Ltd. along with all the intellectual property rights of the consultants reports and information in 2005 after which the neighbouring agricultural land was also brought in to the Shipton development.

New Settlement Master Plan

Under Kilmartin's ownership considerable detailed work was undertaken in respect of the design of a new settlement at Shipton quarry together with the accompanying public transport improvements utilising the existing railway line and bus options to enable its promotion with the Cherwell District Council, Oxfordshire County Council and others. A Master Plan was prepared by architects Feilden Clegg Bradley indicating an overall capacity of around 2,500 dwellings together with 10 hectares of Class B1 employment land capable of producing about 1,500 new jobs. The new settlement was to be accompanied by a primary school together with a significant retail centre which would also provide a variety of community uses and facilities.

The railway line offers the opportunity of an 8-minute journey to Oxford railway station which itself is well placed in terms of access to central area jobs. Detailed discussions were held with the Strategic Rail Authority (before its disbandment) and with Network Rail and railway operators. Bride Parks put forward proposals for a new station to the SRA/Network Rail in the context of the new settlement option.

Further work was undertaken by the architects, Feilden Clegg Bradley, to increase densities and achieve a minimum dwelling number of 3,000 houses. The provisional Master Plan was presented to the Commission for Architect in the Built Environment (CABE).

The promotion of the site for housing was not successful and ownership passed from Kilmartin Group to Earthline Ltd. during 2010.